T0206157

Soil Nailing

Applied Geotechnics series
William Powrie (ed.)

For more information about this series, please visit: https://www.routledge.com/Applied-Geotechnics/book-series/APPGEOT

Soil Nailing
A Practical Guide

Raymond Cheung and Ken Ho

CRC Press
Taylor & Francis Group
Boca Raton London New York

CRC Press is an imprint of the
Taylor & Francis Group, an **informa** business

First edition published 2021
by CRC Press
6000 Broken Sound Parkway NW, Suite 300, Boca Raton, FL 33487-2742

and by CRC Press
2 Park Square, Milton Park, Abingdon, Oxon, OX14 4RN

© 2021 Taylor & Francis Group, LLC
CRC Press is an imprint of Taylor & Francis Group, LLC

Reasonable efforts have been made to publish reliable data and information, but the author and publisher cannot assume responsibility for the validity of all materials or the consequences of their use. The authors and publishers have attempted to trace the copyright holders of all material reproduced in this publication and apologize to copyright holders if permission to publish in this form has not been obtained. If any copyright material has not been acknowledged please write and let us know so we may rectify in any future reprint.

Except as permitted under U.S. Copyright Law, no part of this book may be reprinted, reproduced, transmitted, or utilized in any form by any electronic, mechanical, or other means, now known or hereafter invented, including photocopying, microfilming, and recording, or in any information storage or retrieval system, without written permission from the publishers.

For permission to photocopy or use material electronically from this work, access www.copyright .com or contact the Copyright Clearance Center, Inc. (CCC), 222 Rosewood Drive, Danvers, MA 01923, 978-750-8400. For works that are not available on CCC please contact mpkbookspermissions@tandf.co.uk

Trademark notice: Product or corporate names may be trademarks or registered trademarks, and are used only for identification and explanation without intent to infringe.

Library of Congress Cataloging-in-Publication Data

Names: Cheung, Raymond, (Geotechnical engineer), author. | Ho, Ken, author.
Title: Soil nailing : a practical guide / Raymond Cheung and Ken Ho.
Description: First edition. | Boca Raton : CRC Press, 2021. | Includes bibliographical references and index.
Identifiers: LCCN 2020049302 (print) | LCCN 2020049303 (ebook) | ISBN 9781138031876 (hardback) | ISBN 9780367816261 (ebook)
Subjects: LCSH: Soil nailing.
Classification: LCC TA749 .C46 2021 (print) | LCC TA749 (ebook) | DDC 624.1/51363--dc23
LC record available at https://lccn.loc.gov/2020049302
LC ebook record available at https://lccn.loc.gov/2020049303

ISBN: 978-1-138-03187-6 (hbk)
ISBN: 978-0-367-81626-1 (ebk)

Typeset in Sabon
by Deanta Global Publishing Services, Chennai, India

Contents

List of figures

List of tables

Preface

Soil nailing is an engineering technique that basically combines soil reinforcement and nail heads and/or a facing to stabilize *in situ* soil mass around the world. Since its introduction to the construction industry in the early 1970s, soil nailing has been proven to be a robust, reliable, and cost-effective method. The track record of soil nailing has been excellent in that no major collapses have been reported in properly designed and well-constructed soil nailed structures so far. Nowadays, the technique is commonly used for enhancing the stability of earth structures including slopes, retaining walls, embankments, and excavations. Considerable experience and knowledge of the technique have been gained in the past few decades through systematic technical development work, design, and construction practices.

This book provides a general treatment of the subject of soil nailing, which consolidates the experience and advances made in the development and use of the technique. The book is not exhaustive and is intended for use by postgraduate students, researchers, and practicing civil and geotechnical engineers, who wish to have a more in-depth and fundamental understanding of the theory and practice behind the technique. It presents the basic principles of the technique as well as state-of-the-art knowledge and recommended standard of good practice in respect of design, construction, monitoring, and maintenance of soil nailed structures. Although the book covers a certain amount of theoretical subjects in some depth, the main focus is on the practical aspects which encourages a wider adoption of the technique by practitioners.

During the preparation and development of the book, the authors are indebted in many ways to colleagues who had to endure imperfect versions of the material. We would like to thank Mr. Herman Shiu and Dr. Dominic Lo for numerous discussions and constructive suggestions; to Mr. Edward Chu for helping with the preparation of the worked examples and figures of the book. Figures and tables from journals, proceedings, and books are reproduced with permission from the respective publishers.

Raymond Cheung
Ken Ho
September 2020

Author biographies

Raymond Cheung is Head of the Geotechnical Engineering Office of the Civil Engineering and Development Department, Government of the Hong Kong Special Administrative Region, China. He was involved in the publication of Geoguide 7 (Guide to Soil Nail Design and Construction), which is currently the well-received standard for soil nailing construction in Hong Kong as well as in other Asian countries. Apart from soil nailing, Dr. Cheung's research interest includes reliability and probabilistic analysis of slope, quantitative risk assessment, use of artificial intelligence in slope engineering.

Ken Ho is Deputy Head of the Geotechnical Engineering Office of the Civil Engineering and Development Department, Government of the Hong Kong Special Administrative Region, China. He is an Adjunct Professor of The University of Hong Kong. He has published over 130 papers, including more than ten keynote papers at major international conferences. He delivered the 29th Annual Professor Chin Fung Kee Memorial Lecture at The Institution of Engineers, Malaysia in 2019. Professor Ho is a core member of the Joint Technical Committee (JTC1) on Natural Slopes and Landslides of the Federation of International Geo-engineering Societies (FedIGS).

Symbols

A	Corrosiveness index related to corrosivity assessment of the soil
$A_{s,norm}$	Nominal cross-sectional area of reinforcement
A_s	Area of soil shearing plane
a	Empirical coefficient for calculation of normal stress acting on soil nail
C	Corrosiveness index related to type of soil nailed structures
C	Covariance matrix
c_k'	Factored effective cohesion of soil
c_u	Undrained shear strength of soil
c_a	Adhesion between soil nail and soil
c'	Effective cohesion of soil
D	Diameter of reinforcement/soil nail
D_{max}	Maximum excavation depth
d	Required diameter of reinforcement without considering corrosion
d_{req}	Diameter of reinforcement required taken into corrosion
E	Young's modulus of soil
EA	Axial stiffness of facing
EI	Flexural stiffness of facing
e	Void ratio of soil
FS	Factor of safety
F_B	Factor of safety against bearing failure
F_{FF}	Factor of safety against flexural failure of wall facing
F_G	Factor of safety against global or overall failure of soil nailed structures
F_P	Factor of safety against pullout failure at soil-grout interface
F_{PF}	Factor of safety against punching failure of wall facing
F_R	Factor of safety against pullout failure at grout-reinforcement interface
F_S	Factor of safety against sliding failure of soil nailed structures
F_T	Factor of safety against tensile failure of soil nail reinforcement
f_b	Factor on angle of shearing resistance of soil for calculation of ultimate pullout resistance

f_c	Factor on effective cohesion of soil for calculation of ultimate pullout resistance
f_{cu}	Characteristic strength of cement grout
f_y	Yield strength of steel reinforcement
f_{yk}	Characteristic yield strength of reinforcement
f_ϕ	Factor on angle of shearing resistance of soil for calculation of ultimate pullout resistance
G	Performance function
H	Height of structure/wall
h	Depth of overburden directly above the point in question
K	Constant to calculate depth of general corrosion or pit depth
k_r	Factor relating the average radial effective stress around the nail to the vertical
I	Overall corrosivity index
L	Load
L	Bond length of soil nail
L	Distance between pulse generator and point of discontinuity or mismatch
L, l	Length of soil nail
L_U	Free length of soil nail
L_G	Grouted length of soil nail
LI	Liquidity Index of soil
M	Bending moment on reinforcement / soil nail
M_p	Plastic moment capacity
$M_{driving}$	Driving moment
$M_{resisting}$	Resisting moment
N	Standard penetration resistance
N	Tensile force of soil nail
N_1, N_2	Normal force on the base of wedge 1 or 2
N_{12}, N_{21}	Normal force on inter-wedge boundary
n	Constant to calculate depth of general corrosion or pit depth
P	Probability
P_c	Perimeter of soil nail
P_f	Probability of failure
P_r	Effective perimeter of reinforcement
P_s	Shear force on reinforcement / soil nail
PI	Plasticity index of soil
p	Grouting pressure for calculation of normal stress acting on soil nail
pH	Value of acidity of an aqueous solution
Q_{di}	Load for the ith type of dead load
Q_{ei}	Load for the ith type of extreme loads
Q_{ti}	Load for the ith type of transient loads
q	Surcharge

q_p	Permanent surcharge
q_v	Variable surcharge
q_1, q_2	Surcharge on wedge 1 or 2
R	Resistance
R_{td}	Design tensile force of reinforcement
R_1, R_2	Tangential force on the base of wedge 1 or 2
R_{12}, R_{21}	Tangential force on inter-wedge boundary
r_u	Pore pressure parameter
T	Reinforcement force acting on wedge
T	Design load of soil nail
T	Tensile force of reinforcement
T_a	Initial load for a pullout test
T_d	Design pullout resistance at soil-grout interface
T_c	Creep test load for a creep test
T_{DL1}, T_{DL2}	Intermediate loads for a pullout test
T_L	Ultimate pullout resistance
T_p	Plastic axial force in soil nail
T_p	90% yield load of reinforcement for a pullout test
T_P	Allowable pullout resistance at soil-grout interface
T_R	Allowable pullout resistance at grout-reinforcement interface
T_T	Allowable tensile force of reinforcement
T_{ult}	Ultimate ground-grout bond load of a pullout test
t	Time
u	Pore water pressure
V_i	Peak voltage of incident pulse
V_r	Peak voltage of reflected pulse
v_c	Speed of light in vacuum
v_p	Pulse propagation velocity
W	Self-weight of soil
w	Size of soil nail head
w	Water content of soil
w_1, w_2	Weight of wedge 1 or 2
X	Depth of general corrosion or pit depth at time t
X	Corrosion loss in thickness during the service life of the soil nailed structure
X, x	Vector of design random variables
Z	Electrical impedance at a point of reflection
Z_o	Characteristic electrical impedance
z	Distance measured from origin in the Z direction
α, α_S	Inclination of reinforcement / soil nail
α	Adhesion factor of soil for calculation of ultimate pullout resistance
β	Coefficient of friction at grout-reinforcement interface
β	Reliability index

β_s	Angle of slope/wall
Δe	Metal loss in thickness due to corrosion
δ	Angle of friction between soil nail and soil
δ_h	Horizontal displacement at top of wall facing
$\delta_{h,max}$	Maximum horizontal displacement
δ_o	Influence zone behind the wall face where the ground deformation is negligible
δ_v	Vertical displacement at top of wall facing
ε	Dielectric constant of a material
ε_1	Major principal shear strain of soil
ε_3	Minor principal shear strain of soil
$\Phi()$	Cumulative normal distribution function
Φ	Resistance factor
Φ'	Angle of shearing resistance of soil
Φ_k'	Factored angle of shearing resistance of soil
Γ	Reflection coefficient
γ	Unit weight of soil
γ_{cu}	Partial factor for undrained shear strength
γ_c'	Partial factor for effective cohesion
γ_{di}	Load factor corresponding to dead loads, Q_{di}
γ_{ei}	Load factor corresponding to extreme loads, Q_{ei}
γ_g	Partial factor for self-weight of soil
γ_{qp}	Partial factor for permanent surcharge
γ_{qv}	Partial factor for variable surcharge
γ_s	Partial factor for tensile strength of soil nail
γ_{sd}	Model factor
$\gamma_{tan\phi'}$	Partial factor for tangent of angle of shearing resistance
γ_{ti}	Load factor corresponding to transient loads, Q_{ti}
γ_u	Partial factor for groundwater pressure
γ_γ	Partial factor for unit weight
$\gamma_{\tau b}$	Partial factor for bond stress of soil nail
κ	Coefficient for estimation of horizontal influence zone behind the soil nailed zone
λ_c	Factor on effective cohesion of soil for calculation of ultimate pullout resistance
λ_f	Interface factor
λ_ϕ	Factor on angle of shearing resistance of soil for calculation of ultimate pullout resistance
μ	Vector of mean values
$\mu_{c'}$	Mean of effective cohesion
μ_{FS}	Mean value of factor of safety
$\mu_{\phi'}$	Mean of angle of shearing resistance
μ^*	Apparent coefficient of friction between soil nail and soil
θ	Orientation of reinforcement

θ	Orientation of soil nail with respect to potential failure surface
θ	Angle between soil nail / reinforcement and the normal to potential shear surface
σ_{FS}	Standard deviation of factor of safety
σ_c'	Standard deviation of effective cohesion
σ_{yy}	Vertical stress on shear plane
$\sigma_{\phi'}$	Standard deviation of angle of shearing resistance
σ_N'	Effective normal stress acting on soil nail
σ'_b	Limiting soil bearing stress
σ'_l	Lateral stress on reinforcement / soil nail
σ_v'	Vertical stress acting on soil nail
ν	Poisson ratio of soil
Δ_{FS}	Increase in factor of safety
$\Delta\tau$	Increase of shearing strength of soil
τ_b	Bond stress of soil nail
τ_{EXT}	Extra shearing resistance due to reinforcement
τ_{ult}	Ultimate bond stress of soil nail

Chapter 1

Introduction

1.1 OVERVIEW AND INTENDED READERSHIP OF THE BOOK

Soil nailing is one of the engineering techniques commonly employed to stabilize earth structures around the world. The technique is used to reinforce *in situ* soil mass, either natural or deposited by human action, with closely spaced, relatively slender unstressed tension-carrying structural elements. These elements, which are called soil nails, comprise steel or other engineering materials such as fiber reinforced polymer. Soil nails can be either driven or fired directly into the ground, or more commonly boreholes are drilled followed by insertion of reinforcement and grouting up of the annulus. Under normal circumstances, soil nails act as passive elements and do not carry any stresses upon installation. When there is a change in the stress-strain state of the reinforced soil mass and hence ground deformation, tensile forces will be developed in the soil nails through the frictional interaction between the soil nails and the ground. Soil nailing did not gain popularity until the 1970s when engineers started to realize that the technique could offer an effective, robust, and economical reinforcing system for a variety of ground conditions. The construction process generally causes minimal nuisance, both in terms of vibration and noise, to the environment. The maintenance effort required throughout the lifetime of a soil nailed structure is also minimal. More importantly, the track record has been excellent in that no major collapses have been reported in properly designed and well-constructed soil nailed structures so far. Nowadays, the technique is commonly used for enhancing the stability of earth structures including slopes, retaining walls, embankments, and excavations. Considerable experience and knowledge of the technique have been gained in the past few decades through systematic technical development work, comprising laboratory tests, numerical modeling, physical modeling, and site trials, and covering design and construction practices.

This book consolidates the experience and advances made in the development and use of the soil nailing technique and encourages a wider adoption of the technique by practitioners. The book is intended for use by

1

postgraduate students, researchers, and practicing civil and geotechnical engineers, who wish to have a more in-depth and fundamental understanding of the theory and practice behind the technique. It presents the basic principles of the technique as well as state-of-the-art knowledge and recommended standard of good practice in respect of design, construction, monitoring, and maintenance of soil nailed structures. Selected projects and relevant studies are also highlighted in order to illustrate the innovative use of the soil nailing technique from the perspective of design, construction, and quality control.

1.2 EARTH STRUCTURE STABILIZATION METHODS

1.2.1 General

A number of methods can be used to stabilize earth structures. These methods are developed upon two basic engineering principles, namely (i) reduction of driving forces and (ii) enhancement of resistance. In selecting and designing a suitable method for earth structure stabilization, one should consider the merits and limitations of the options in respect of geotechnical, construction, environmental, maintenance, and financial implications. Table 1.1 summarizes various common earth structure stabilization methods that are adopted nowadays in civil engineering projects.

1.2.2 Reduction of driving forces

A simple way to improve the stability of an earth structure is to directly reduce the driving forces, which are mostly gravitational in nature. This can be achieved by ground re-profiling, surface treatment, or provision of a drainage system. The three approaches may be adopted in combination for different types of earth structures. Various methods of ground re-profiling, including grading the earth structure into a flatter gradient, benching the earth structure, trimming over-steepened sections, etc., are commonly employed in geotechnical practice.

Surface treatment is also a common measure to increase the stability of an earth structure. The treatment may be in the form of hard, flexible, or soft (i.e. vegetation) cover. The main purpose of surface treatment is to minimize surface runoff from entering a slope, embankment, or excavation as well as to prevent surface erosion. Materials such as shotcrete (sprayed concrete), stone pitching, and masonry are commonly used for hard surface cover, whereas flexible facing typically comprises metallic netting or meshing, sometimes with pre-tension. Soft surface cover may comprise vegetation such as hydroseeding and hydro-mulching, or bioengineering techniques, which entail the use of living plant materials, frequently in association with inert materials such as rock, wood, geosynthetics, and geocomposite. Plant

Table 1.1 Summary of common earth structure stabilization methods

Principle	Treatment		Remarks
Reduce Driving Forces	Re-profile	Grading Trimming Benching	Reduction of dead weight
	Surface Treatment	Hard facing Flexible facing Soft (vegetation) facing	Prevention or reduction of water ingress into the ground mass and minimization of erosion
	Drainage	Surface Subsurface	
Enhance Resistance	External (Active)	Retaining walls Piles Buttresses Counterweight fills Prestressed ground anchors	Provision of resistance through external structural elements
	Internal (Passive)	Reinforcement Grouting Compaction Drainage	Provision of resistance through internal structural elements or increase in shearing resistance of the ground mass

materials on a sloping surface provide protection against surface erosion by reinforcing the soil mass at shallow level through root action. Their roots can also reduce pore water pressures within the slope by intercepting rainfall and enhancing evapotranspiration.

Water control is another effective measure to improve the stability of an earth structure. The primary sources of water in an earth structure are from surface runoff and groundwater. Water control is generally undertaken through provision of surface drainage and subsurface drainage. A properly functioning drainage system can also both reduce the weight of the soil mass tending to cause instability and increase the shear strength of soil in the earth structure.

1.2.3 Enhancement of resistance

Resistance against instability of the earth structure can be provided by external or internal means. External means are those active engineered structures constructed to resist lateral forces imposed by earth pressures and pore water pressures. Examples of external means include retaining walls, piles, buttresses, prestressed ground anchors, etc. Internal means, on the other hand, are passive engineered structures where the stabilization forces are mobilized only when ground movement occurs or where the soil shear strength is improved by chemical or mechanical means. Examples of the former include soil nails, soil dowels, and reinforced earth, whereas grouting and compaction are examples of the latter.

1.2.4 Difference between soil nailing and other soil reinforcement methods

As indicated in Table 1.1, soil reinforcement is one of the major construction methods commonly employed to stabilize earth structures. Although soil nailing shares some common features with other soil reinforcement methods like soil dowels, reinforced earth, and prestressed ground anchors, there are fundamental differences between them. A clear distinction between soil nails and soil dowels is their primary mechanical function. The principal resistance of soil nails primarily comes from their tensile capacity, with shearing and bending as a secondary benefit. In contrast, soil dowels provide shearing and bending resistance with tension as a secondary benefit only. In general, soil doweling is applied to reduce or arrest movement of slopes along well-defined shear surfaces and slopes treated by soil dowels are usually much flatter than those for soil nailing.

Reinforced earth is another common method for reinforcing new earth embankments. While reinforced earth is a "bottom-up" technique, soil nailing is a technique which can either be "bottom-up" (e.g. reinforcing existing slopes) or "top-down" (e.g. reinforcing excavations). This leads to a different distribution of forces developed in the reinforcement between the two techniques. The maximum deformation of an earth structure with soil nailing commonly occurs at the top of the wall, whereas that for a reinforced earth structure occurs at the toe instead. Also, grouting techniques are usually employed in soil nailing to bond the reinforcement to the ground. In reinforced earth, however, friction is generated directly at the interface of the reinforcement strip and the ground. As soil nailing is an *in situ* reinforcement method exploiting natural ground, the properties of the soils cannot be selected or controlled as in the case of reinforced earth.

For prestressed ground anchors, only a limited portion of the anchors is connected to the ground directly, whereas the entire length of the soil nails is connected to the ground. This gives rise to different stress distributions in the soil mass. Ground anchors tend to be longer (typically about 15 m to 40 m) than soil nails (typically about 5 m to 20 m) and are usually prestressed to high tension. Soil nails, on the other hand, are basically passive elements in which tension is mobilized only where there is ground deformation. Thus, ground anchors prevent movement of the earth structures, whereas soil nails allow a certain degree of soil deformation in order to mobilize tensile resistance. The magnitude of the tension in prestressed ground anchors (typically 500 to 1,000 kN) is usually much higher than that mobilized in soil nails (typically 50 to 200 kN). Because of this, substantial bearing arrangement must be provided at the anchor heads to minimize potential punching failure. Furthermore, soil nails are commonly installed at a much higher density and as a result, the consequence of a unit failure is not as severe as prestressed ground anchor. Generally speaking, prestressed ground anchors are suitable for potential deep-seated failures or slopes with pre-existing weak planes, whereas soil nails are suitable for steep excavations and improvement works to existing slopes, retaining walls, and embankments with the potential for shallow instabilities.

1.3 A BRIEF HISTORY OF THE SOIL NAILING TECHNIQUE

1.3.1 Origins of the soil nailing technique

The soil nailing technique evolved in the early 1960s, partly from rock bolting and multi-anchorage systems, and partly from reinforced fill (Clouterre, 1991; Byrne et al., 1998). The New Austrian Tunneling Method (NATM) introduced in the early 1960s by Rabcewicz (1964a & b, 1965) was the premier prototype to use steel bars and shotcrete (sprayed concrete) to reinforce the ground (Figure 1.1). The design intent is to install and grout the steel bars immediately after completion of tunnel excavation and meanwhile sprayed concrete with a steel wire mesh is applied to the tunnel face to develop a flexible support system in order to control ground deformation in underground excavations. Unlike the conventional tunneling technique whereby a rigid lining is employed to prevent soil deformation and hence the lining would be subjected to full loading from the ground and water pressures, NATM enables an active zone to be formed in the vicinity of the excavation and the lining is consequently subjected to reduced loading. With this new method, a thinner tunnel lining than the conventional one can be achieved as the tunnel support. This new technique was initially used in hard rock tunneling projects in Germany in the early 1960s, and the application was then extended to less competent ground materials such as graphitic shales in the Massenberg Tunnel and Keuper Marl in the Schwaikhem Tunnel. The innovative idea of NATM inspired the development of the soil nailing technique. In the 1970s, engineers in France,

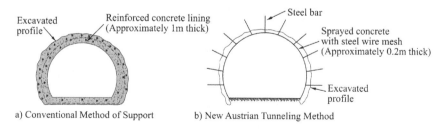

Figure 1.1 Schematic comparison of the New Austrian Tunneling Method (NATM) and Conventional Supporting Method.

Germany, and North America started to exploit independently the use of the soil nailing technique for stabilizing slopes and excavations. Since the mid-1970s, the technique has been researched fairly extensively by various collaborations of practitioners, universities, and government agencies.

1.3.2 Development of the soil nailing technique in France and Germany

The first recorded use of soil nailing in Europe was in 1972, when the French Railways undertook the stabilization of an 18 m high battered face in cemented Fontainebleau sand. The French contractor Bouygues gained experience from NATM and considered that a similar technique could be applied as a temporary support for slopes comprising soft rock or soil. In 1972, Bouygues, in Joint Venture with the specialist contractor, Soletanche, used soil nailing on a 70° cut slope in cemented sand for a railway widening project near Versailles. As illustrated in Figure 1.2, two additional railway tracks were accommodated by forming the 70o soil nailed slope. A total of 12,000 m² of slope face was stabilized by over 25,000 steel bars grouted into predrilled holes up to 6 m long. The successful application of soil nailing in the project encouraged others and the technique was subsequently adapted for use in Paris underground extensions in 1974 wherein the nails were driven, rather than grouted, into the ground. Since then, soil nailing has gained popularity in France, and a number of specialist contractors started to develop their own installation methods based on drilling and grouting. At that time, the technique was used primarily for temporary structures because of the lack of knowledge about the long-term behavior of soil nailed structures. Indeed, this new technique had not been

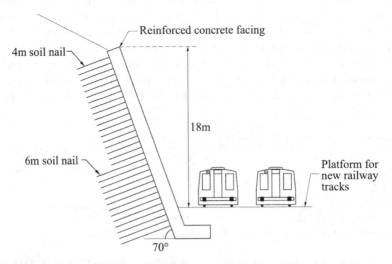

Figure 1.2 Section of a soil nailed wall near Versailles, France.

scientifically and systematically studied in France until the late 1980s when a large research program, called "*Clouterre*," was carried out to study the behavior of soil nailed retaining walls during construction, in service, and at failure.

According to Gässler (1997), after some successful applications of NATM in less competent materials like silts, sand, and gravel, the idea of *in situ* reinforced soil was taken up in Germany in the early 1970s. The first application of soil nailing in Germany involved the construction of a retaining wall in 1975. By the late 1970s, the technique had been applied to some 20 engineering projects (Gässler & Gudehus, 1981). The first systematic research program on soil nailing in the world, namely "*Bodenvernagelung*," which involved both theoretical stability analyses, physical model tests, and seven instrumented full-scale field tests on retaining walls (up to 6.9 m high), was carried out in Germany between 1975 and 1981 by the University of Karlsruhe and the specialist contractor Karl Bauer AG (Gässler, 1997; Byrne et al., 1998). The terminology of "soil nailing" is derived from two German words, namely "Boden" (literally means soil) and "Vernagelung" (literally means nailing). The major findings of the study indicate that the reinforced soil mass acted like a gravity wall and it was capable of withstanding dynamic loads. The nail length was typically 0.5 to 0.8 times the wall height and nail spacing was about 1.5 m. Also, the earth pressure on the wall facing was found to be about 40% to 70% of the active soil pressures, and the soil nails did not undergo significant bending until the onset of failure. The design and construction approaches as developed in the above research program subsequently became the code of practice for soil nailing in Germany. Nonetheless, as soil nailing was still considered to be a developing technique in the 1980s, it was not included in the German standard (DIN). Figure 1.3 shows the first sketch of the construction sequence of soil nailing in its original form at the International Conference on Soil Reinforcement in Paris in 1979.

Subsequent development work was initiated independently in France and the United States, particularly the 4-year systematic research program "*Clouterre*" (*cloutage du sol* means soil nailing in French) launched by the French Minister of Transport in 1986. The Clouterre program involved three large-scale experiments in fill (Fontainebleau sand) and the monitoring of six full-scale in-service soil nailed structures. The study covered the entire design and construction process, from geotechnical investigation, design, construction, and quality control. The major findings of the research suggested that the principal resistance provided by the soil nails during excavation was tension, whereas bending of nails was not observed until the occurrence of large deformation. The maximum tensile force was not located at the wall face but tended to be close to the failure zone. Also, the horizonal and vertical displacements of the nailed structure were of the same magnitude and the largest displacement at the wall top was found to be about 0.1% to 0.3% of the wall height. The research and development

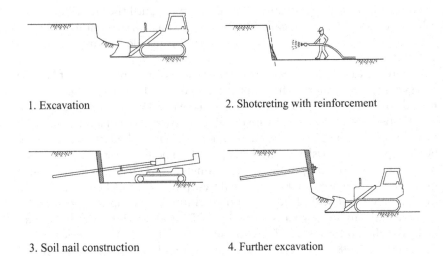

1. Excavation   2. Shotcreting with reinforcement

3. Soil nail construction   4. Further excavation

Figure 1.3 Construction sequence of soil nailing. (Modified Stocker et al., 1979.)

work formed the basis for the formulation of the design approach and construction practice in France, which subsequently led to the publication of the French guidelines on soil nailing (Clouterre, 1991).

1.3.3 Development of the soil nailing technique in the UK

A comprehensive summary of the early development of the soil nailing technique in the UK is given by Barley (1992) and Phear et al. (2005). In comparison with that in France and Germany, the development of the technique in the UK has been relatively slow. The main reasons for this were probably the concern about the long-term durability of soil nails and the role of their shear resistance and bending resistance when compared with tension in stabilizing slopes and retaining structures. Research and development work on soil nailing began in the early 1990s, principally by the universities, particularly the University of Oxford, the University of Wales at Cardiff, Imperial College of Science, Technology and Medicine, London, and the University of Dundee. Other research institutions such as the Transport Research Laboratory also took part (Phear et al., 2005). The focus of the work at that time included the role of bending and shear relative to tensile resistance in soil nails, nail-ground interaction through large shear box tests, soil nail pullout resistance, durability, etc. Unlike the practice in France and Germany, where soil nails have been used for a variety of applications, they have been used in the UK mostly for stabilizing existing slopes, retaining walls, and embankments. Up to the mid-1990s, the UK soil nailing market was still very limited (Perry et al., 2001;

Phear et al., 2005). The standard of good practice promulgated specifically for design of soil nails in the early 1990s was the DOT Advice Note, HA 68/94 (DOT, 1994), which was published by the Highways Agency of the Department of Transport in 1994. The latest standards of good practice on use of soil nailing, which are currently in force in the UK, include CIRIA Report No. C637 (Phear et al., 2005), BS EN 14490 (BSI, 2010), BS 8006-2 (BSI, 2013), and O'Donovan et al. (2020). According to O'Donovan et al. (2020), the use of soil nailing in the UK has become popular since the mid-2000s. One of the reasons for this could be the advent of the self-drilled hollow bar soil nailing technique, which enables the drilling, placement of nail reinforcement and injection of cement grout to be combined as a single operation.

1.3.4 Development of the soil nailing technique in North America

In North America, the majority use of soil nailing has been for temporary support to excavations, in particular for building construction in the urban area. According to Byrne et al. (1998), the soil nailing technique has been applied to wall construction in Canada, the United States, and Mexico as early as the late 1960s to early 1970s. However, the first documented application of soil nailing in North America was in Vancouver, BC, Canada during the early 1970s in connection with a temporary excavation. At that time, soil nailing had been applied mainly to temporary excavations in western Canada. Meanwhile, the first documented application of the technique in the United States was in 1976 when soil nails were used to support a 13.7 m deep excavation in dense silty sand for basement construction in Portland, Oregon. Researchers of the University of California at Davis were involved in the instrumentation of this excavation and as follow-up work, a full-scale nailed wall was constructed and tested at the university campus in 1979. This formed part of the research program on soil nailing spanning from 1976 to 1981, which was called "lateral earth support system" in the United States (Shen et al., 1981a & b). The results of the research led to the development of the limit equilibrium analysis method, which is commonly referred to as the "Davis method." Since then, the use of soil nailing in retaining structures has increased considerably for road projects. The Federal Highway Administration (FHWA) in the United States initiated a series of research projects from the late 1980s to early 1990s with a view to developing a standard of good practice for soil nail wall construction (Byrne et al., 1998). For example, FHWA commissioned the first comprehensive study of soil nailing in 1989, which resulted in the publication of the first document for design and construction of soil nails (Elias & Juran, 1991). The latest guidance documents on soil nail walls were in the form of a Manual (Byrne et al., 1998) and Geotechnical Engineering Circular (Lazarte et al., 2015). Soil nailing technology has become very popular in North America. This is attributed to its cost effectiveness, less demand for construction equipment, and its high structural redundancy (a large number of reinforcing elements per unit area of wall as compared to other systems).

1.3.5 Development of the soil nailing technique in South America

The early development and experience of the soil nailing technique in South America is best represented by Brazil (Ortigao et al., 1995). NATM was first applied by tunneling contractors at the Cantareira water supply project in Brazil in 1970. The crown and face of the tunnel were reinforced by bolting or nailing with a flexible shotcrete layer. The same technique was also applied to stabilize the slopes of the tunnel portal in the same project. It was named *"solo enraizado"* in Portuguese, which is equivalent to soil rooting. Shortly afterward, the technique was applied to the construction of several other tunnel portals and slopes. However, the experience was not properly recorded and reported. The information was only available to a few engineers involved in the work. Nevertheless, the use of soil nailing in Brazil can be divided into two phases. The first phase began in the early 1970s when the technique was used empirically based on NATM experience in a prescriptive manner (i.e. without calculations), whereas the second phase began in the 1980s when design procedures based on limit equilibrium evolved, enabling sizable geotechnical structures to be designed based on recognized method of analysis.

1.3.6 Development of the soil nailing technique in Japan

The first application of the soil nailing technique in Japan was to reinforce the roadside slopes of Yagiyama bypass in Kyushu in 1982 (Tayama & Kawai, 1999). Since then, relevant technical development work including laboratory tests and field trials was conducted in the 1980s (e.g. Tatsuoka et al., 1983; Mori & Asahi, 1986). In 1987, a technical paper on the design and construction of soil nailed slopes was published by the Japan Highway Public Corporation (JHPC, 1987), which became the first guidance document for application of soil nailing in Japan. In 1998, a more comprehensive guidance document, entitled Design and Construction Guidelines for Soil Nailing (JHPC, 1998), was issued and became the state-of-practice for application of the soil nailing technique in Japan. This document covers planning, design, construction, testing, and maintenance of soil nailed structures, particularly for roadside slopes. According to Hirano (2003), about 58% of the soil nailing application in Japan involved preventive and remedial works for existing slopes, whereas the remaining 42% was related to excavations. Among the excavations, about 40% were steep slope cuttings with a gradient greater than 1:0.5 (about 63°).

1.3.7 Development of the soil nailing technique in Hong Kong

The soil nailing technique was introduced to Hong Kong in the 1980s. The technique was first used as a prescriptive method to provide support to soil cuts in deeply weathered rocks. This was followed by a few cases where soil nails acting as passive anchors or tieback systems were used in excavations. Some of the impetus for these early cases stemmed from the desire to find an alternative to prestressed ground anchors, which require long-term monitoring. In the mid-1980s, a small number of soil nailed supports to temporary cuts were made. In the early 1990s, the experience of design and construction of soil nails in upgrading existing substandard cut slopes and retaining walls was summarized by Powell & Watkins (1990), which became the state-of-practice in Hong Kong at that time. The soil nailing technique has gained popularity since the mid-1990s and it has been applied to thousands of slopes and retaining walls both in the government sector and private sector. Experience gained through applications and insights from studies and field monitoring results over the years led to the development of the technique in respect of analysis method, design approach, construction, quality control, and maintenance. In recent years, the Geotechnical Engineering Office (GEO), a specialist arm of the Civil Engineering and Development Department of the Hong Kong Special Administrative Region Government, has embarked, with participation from local research institutes and practitioners, on a program of systematic soil nailing related studies, which included literature reviews, field tests, laboratory investigations, physical and numerical modeling. To consolidate the technical development work and practical experience over the past decades, the GEO published a guidance document, Geoguide 7—Guide to Soil Nail Design and Construction (GEO, 2008), to promulgate the recommended standard of good practice for the design, construction, monitoring, and maintenance of soil nailed structures. To date, soil nailing is the most common earth structure stabilizing method in Hong Kong, particularly for slopes, retaining walls, and excavations. Some 5,000 slopes and retaining walls have been upgraded using soil nails.

1.4 COMPONENTS OF A SOIL NAILED STRUCTURE

Soil nails can be installed into the ground by various methods such as driving, firing, and drilling-and-grouting (see Chapter 7 for more details). The vast majority of the soil nails around the world are installed by means of the

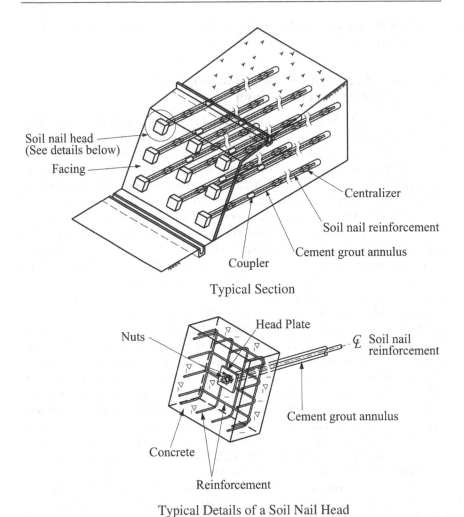

Typical Section

Typical Details of a Soil Nail Head

Figure 1.4 Schematic diagram of a soil nailed structure.

drill-and-grout method. For illustration purposes, Figure 1.4 shows schematically a typical soil nailed structure using this installation method. The major components of the structure are summarized in Table 1.2.

1.5 AREAS OF APPLICATION

With the technical development work and successful application of soil nailing over the past few decades, the technique has evolved and proved to be highly effective, economical, and robust in slope stabilization and

Table 1.2 Summary of the major components of a soil nailed structure

Component	Description
Soil Nail Reinforcement (Tendon)	This is the main structural element of a soil nailed structure. Its primary function is to provide tensile resistance along the length of a soil nail. The reinforcement is typically a solid steel bar or rod, made of either mild steel or high-tensile steel. Other types of materials, such as fiber reinforced polymer, can also be used as the soil nail reinforcement. In some countries, such as the United States, hollow or thread soil nail reinforcing bars are also common for soil nailing application. Soil nail reinforcement can be driven or fired into the ground, or more commonly placed in predrilled holes and contained with a suitable cement grout.
Reinforcement Connector (Coupler)	Soil nail reinforcement may be constructed from several sections. Couplers are structural devices used for joining sections of reinforcement together. There are various types of couplers, such as couplers with threads, with shear bolts, and with metal sleeves swaged onto reinforcement.
Centralizer (Spacer)	Centralizers are devices made of non-corrodible materials such as polyvinyl chloride (PVC) and other synthetic materials that are not detrimental to the soil nail reinforcement. They are installed around and along the soil nail reinforcement at regular intervals to ensure a minimum thickness of grout completely covers the reinforcement.
Cement Grout Annulus	Cement grout, made of cement and water, is placed in a predrilled hole after the insertion of a soil nail reinforcement. The cement grout annulus provides intimate contact between a soil nail reinforcement and the ground. It serves the primary function of transferring stresses between the ground and the soil nail reinforcement. It also provides a nominal level of corrosion protection to the reinforcement.
Corrosion Protection Measures	Different types of corrosion protection measures are required to improve durability of steel reinforcement, depending on the design life, and soil and water aggressivity. Common types of corrosion protection measures are direct coatings (e.g. hot-dip galvanizing or fusion-bonded epoxy) and encapsulating the soil nail reinforcement in a continuous corrugated plastic sheathing of high-density polyethylene (HDPE). Sacrificial steel is also a simplest form of corrosion protection measure. Heat-shrinkable sleeves made of polyethylene and anti-corrosion mastic sealant material are commonly used to protect couplers.
Soil Nail Head	A soil nail head typically comprises a reinforced concrete pad, a steel bearing plate, and hexagonal nut. The bearing plate and the nut, which may be embedded within or bedded onto the concrete pad, provide connection between the soil nail reinforcement and the concrete pad. The primary function of a soil nail head is to provide a reaction for the individual soil nail to mobilize its tensile force. It also contributes to the local stability of the ground near the slope surface and between soil nails. Soil nail heads can be recessed into the slope to minimize visual intrusion.

(Continued)

Table 1.2 (Continued) Summary of the major components of a soil nailed structure

Component	Description
Slope Facing	A slope facing generally serves to protect the slope surface and minimize erosion and other adverse effects of surface water on the slope. It may be soft, flexible, hard, or a combination of the three. A soft slope facing is non-structural, whereas a flexible or hard slope facing can be either structural or non-structural. A structural slope facing can enhance the stability of a soil nailed structure by transferring the loads from the free surface in-between the soil nail heads to the soil nails and redistributing the forces between soil nails. The most common type of soft facing is vegetation cover, often in association with an erosion control mat and a steel wire mesh. Some proprietary products of flexible facing are also available. Hard facing includes sprayed concrete, reinforced concrete, and stone pitching. Structural concrete beams and grillages can also be constructed on the slope surface to connect the soil nail heads together in order to promote integral action of the soil nailed structure.

earth retaining purposes. Nowadays, soil nailing has been applied to a wide range of geotechnical structures including stabilization of slopes, retaining walls, embankments, and excavations. Recently, the application of the technique has been extended to preservation of heritage monuments and historic structures (e.g. Kulczykowski et al., 2017). Figure 1.5 illustrates the various common applications of soil nailing in engineering projects.

1.5.1 Slopes

One of the most common areas of application of the soil nailing technique is slope engineering. Soil nailing is used to improve the safety of cuts, fills,

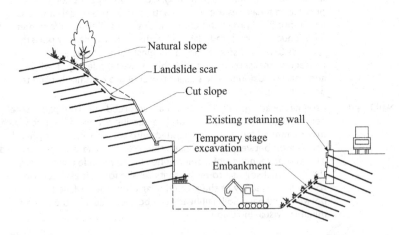

Figure 1.5 Common applications of soil nailing.

retaining walls, and natural slopes where they are deemed to pose an unacceptable risk to adjacent facilities, such as roads and buildings. The technique can be used to stabilize existing slopes or new slopes with a wide range of gradients. Numerous successful cases of application are cited in the literature, such as Bruce & Jewell (1986 & 1987), Ortigao et al. (1995), and Phear et al. (2005).

For cases of new cuts where the ground is excavated at an angle steeper than that at which it can stand safely, soil nailing offers a simple and attractive scheme without increasing land take at the slope crest. These new cuts may be formed in natural ground or existing earthworks. They are normally constructed incrementally where each increment, generally less than three meters high, should consist of a step of excavation to minimize disturbance to the ground as well as to ensure local stability. As excavation proceeds, the soil nails would become progressively loaded. The sequence of excavation should continue until the full cut is achieved.

Unlike the cases in new cut slopes, particularly those with steep gradients, soil nails installed in existing slopes usually remain unloaded unless or until there is movement of the slope caused by, for example, transient load arising from elevations of groundwater level, surcharge, or earthquake. For existing fill slopes, soil nails could be applied as that for cut slopes to improve their stability so long as fill materials do not show contractive or collapsible behavior upon shearing. However, if the fill materials exhibit contractive behavior upon shearing or possess a structure that may lead to strain softening or collapse, soil nailing on its own following conventional application may not be a suitable scheme.

For natural slopes, due consideration should be given to prevent the soil nailing scheme from adversely affecting the natural environment and ecology (including existing vegetation and wildlife). Soft facing systems in association with suitable landscaping measures should be employed whenever possible.

1.5.2 Existing retaining walls and embankments

Soil nailing is also a suitable scheme for strengthening retaining walls and embankments. The technique has been widely used in stabilizing existing retaining walls and embankments as part of roadway or railway projects (e.g. Perry et al., 2001; Phear et al., 2005; Lazarte et al., 2015). It can be applied to various types of retaining walls such as stone, brick, and concrete. Soil nails can be installed directly through the wall face if it is sufficiently stable to resist drilling. However, due consideration should be given to the connection details between soil nail heads and the wall, particularly where the wall is suffering from severe distress. As further ground movement is necessary to mobilize tensile force in soil nails, special measures (such as grouting) may be required to stabilize the ground immediately behind the retaining wall.

Many old embankments, including those for railways, highways, and dams, have been stabilized by soil nails. The reinforced soil mass becomes a homogenous and resistant unit, like a gravity wall, to support the unreinforced ground behind it.

1.5.3 Temporary excavations

Another common area of soil nailing application is to facilitate temporary excavations. Special attention should be given to planning and designing a staged cut where the height of the exposed slope face should be determined on the basis of its stability, particularly before the construction of soil nail heads. If the temporary excavation involves the use of lateral support, soil nails can serve as tiebacks. Where a temporary excavation is carried out in cohesive soils, special care should be exercised with regard to the potential effect of soil creep or swelling due to drainage with time on the stability and serviceability of the excavation, in particular if the soil nails are designed to carry sustained loads. Recently, soil nailing has been used successfully to support a 30 m and 50 m high temporary excavation for construction of the Chinese Embassy in the United States (Bonita et al., 2006) and an underground railway station in Hong Kong (Kwong & Chim, 2015), respectively.

1.5.4 Urgent repair works

Soil nailing is also suitable as urgent repairs to existing distressed slopes, retaining walls, embankments, excavations, and landslide scars. Ongoing ground deformation of the distressed geotechnical features is required to mobilize the reinforcing resistance of the soil nails. In the United States, soil nails have been used to stabilize many masonry or reinforced concrete retaining walls which exhibited excessive deformation or structural deterioration, as well as mechanically stabilized earth (MSE) walls or crib walls that have problems of reinforcement corrosion or poor quality of backfill. In Hong Kong, soil nailing is a very common urgent repair measure for distressed slopes or landslide scars during the rainy season.

1.6 MERITS AND LIMITATIONS

The soil nailing technique offers an attractive engineering solution for stabilizing slopes, retaining walls, embankments, and excavations. However, not all the ground conditions are intrinsically suitable for soil nailing. Some soils may present problems either during construction or in the long term, which make the soil nailing scheme impractical, uneconomical, or even unsafe. The key to the success of a soil nailing scheme hinges on the ability to form an integral reinforced zone by the nails and the soil mass into which they

Table 1.3 Merits and constraints of the soil nailing technique

Merits	Constraints
Flexibility: suitable for cramped sites with difficult access because the construction plant required is relatively light and mobile.	Sterilization of ground: the zone occupied by soil nails is sterilized and the site poses constraints to future development.
Adaptability: easily copes with site constraints and variations in ground conditions encountered during construction by adjusting the number, inclination, and location of soil nails.	Soil creep: not suitable for application in soft clays or sensitive soils subject to creep because nail capacity may not be economically developed, which calls for a very high density and considerable lengths of reinforcement.
Environmental friendliness: compatible with the environment as no major earthworks and tree felling are needed; less disruptive to traffic and causes less adverse environmental impact in terms of noise and vibration.	Corrosion: not suitable for application in highly aggressive soils, particularly where long-term durability is a concern.
Robustness and reliability: as a significant number of soil nails are typically involved, it is less sensitive to undetected adverse geological features and seismic loads than unsupported cuts or other soil reinforcement methods like prestressed ground anchors.	Underground obstructions: presence of utilities, underground structures, or other buried obstructions can pose restrictions to the length and layout of soil nails.
Ductility: failure mode is likely to be ductile, thus providing warning signs before failure; soil nailed walls have also shown to perform well during seismic events due to system ductility.	Groundwater: presence of high groundwater levels may lead to construction difficulties in hole drilling and grouting.
Space: more space can be provided at the bottom of excavation, particularly when compared with conventional braced excavations.	Sites with a history of failures: effectiveness of soil nails may be compromised at sites with past large landslides involving deep-seated failure due to disturbance of the ground.
Economy: track records in Europe and North America indicate a saving of 10% to 30% in cost when compared with other conventional methods such as cutting back and retaining wall construction.	Climate: the performance of soil nails is subject to uncertainty in cold climate where the soil, in particular swelling clays or soils that are sensitive to frost, is subjected to freeze-thaw cycles.
Construction rate: fast construction rate can be achieved easily by adequate drilling equipment; less affected by adverse weather.	Permeability: presence of permeable ground, such as ground with cobbles, boulders, highly fractured rocks, open joints, or voids, can present construction difficulties due to potential grout leakage problems.
	Stratigraphy: presence of cobbles and boulders may preclude the use of driven or fired soil nails.
	Jointing and bedding: presence of adverse jointing or bedding may pose instability problems in the short and long term.
	Long soil nails: long soil nails (e.g. in excess of around 20 m) are difficult to install, and thus the soil nailing technique may not be appropriate for deep-seated landslides or sizable slopes.
	Ground deformation: as the mobilization of soil nail forces will be accompanied by ground deformation (plus displacement) induced by step excavation prior to installation of soil nails, the effects on nearby structures, facilities, or services may have to be considered, particularly in the case of soil nailed excavations.

are installed. A thorough understanding of the behavior of the soil nailing technique under different ground conditions is of the essence. Table 1.3 summarizes the typical merits and constraints of the soil nailing technique.

1.7 CONCLUDING REMARKS

Soil nailing is an *in situ* ground reinforcement technique that involves the installation of closely spaced slender structural elements, commonly known as soil nails, either by driving, firing, or more commonly the drill-and-grout method. A soil nailed earth structure basically includes the *in situ* ground to be reinforced, soil nails, nail heads, and/or facing. The technique improves the stability of slopes, retaining walls, embankments, and excavations principally by transferring loads to the ground through the mobilization of tensile forces in the soil nails. The soil nails act to limit the ground deformation near the exposed facing and transfer the stresses to a more stable zone behind the reinforced soil mass. Soil nails are essentially passive elements in which tension is mobilized only if there is ground deformation. The technique evolved in the early 1970s, partly from the techniques for rock bolting and multi-anchorage systems, and partly from the reinforced fill technique. Subsequent technical development work of soil nailing was carried out independently in Europe, the Americas, and Asia from the mid-1970s to the late 2000s. To date, soil nailing has a good performance record and has become one of the common engineering techniques for stabilizing earth structures around the world. In general, the ground conditions most suited for soil nailing include residual soils, weathered rocks with no unfavorable joint orientation, granular soils, and stiff cohesive soils. However, soils that are poorly graded, highly susceptible to frost, with a high organic content, soft clays that are susceptible to creep, and rocks with open joints may not be suitable for soil nailing. In this chapter, salient aspects of the evolution and history of development of the soil nailing technique, its areas of application, the key components, as well as the merits and constraints of the technique are presented. Soil nailing offers an attractive solution to stabilize earth structures from the perspective of its flexibility and cost-effectiveness. Readers should, however, thoroughly understand the merits and constraints of the technique.

REFERENCES

Barley, A.D. (1992). Soil nailing case histories and developments. Proceedings of the ICE Conference on Retaining Structures, Cambridge, pp. 559–573.

Bonita, G.A., Tarquinio, F.S. & Wagner, L. (2006). Soil nail support of excavation system for the Embassy of the Peoples Republic of China in the United States. In Proceedings of the DFI's 31st Annual Conference on Deep Foundations, Washington, DC.

Bruce, D.A. & Jewell, R.A. (1986). Soil nailing: Application and practice: Part 1, *Ground Engineering*, vol. 19, no. 8, pp. 10–15.

Bruce, D.A. & Jewell, R.A. (1987). Soil nailing: Application and practice: Part 2, *Ground Engineering*, vol. 20, no. 1, pp. 21–33.

BSI (British Standards Institution) (2010). Execution of Special Geotechnical Works: Soil Nailing (BS EN 14490: 2010). British Standards Institution, London, UK, 68 p.

BSI (2013). Code of Practice for Strengthened/Reinforced Soils: Part 2: Soil Nail Design (BS 8006-2: 2011). British Standards Institution, London, UK, 104 p.

Byrne, R.J., Cotton, D., Porterfield, J., Wolschlag, C. & Ueblacker, G. (1998). Manual for Design and Construction Monitoring of Soil Nail Walls. Federal Highway Administration, US Department of Transportation, Washington, DC, USA, Report No. FHWA-SA-96-069R, 530 p.

Clouterre (1991). French National Research Project Clouterre: Recommendations Clouterre (English Translation 1993). Federal Highway Administration, US Department of Transportation, Washington, DC, USA, Report No. FHWA-SA-93-026, 321 p.

DOT (Department of Transport) (1994). Design methods for the reinforcement of highway slopes by reinforced soil and soil nailing techniques. In *Design Manual for Roads and Bridges, Part 4 (HA 68/94)*. The Highways Agency, The Department of Transport, UK, 112 p.

Elias, V. & Juran, I. (1991). Soil nailing for stabilization of highway slopes and excavations. Federal Highway Administration, US Department of Transportation, Washington, DC, USA, Report No. FHWA-RD-89-198, 210 p.

Gässler, G. (1997). Design of reinforced excavations and natural slopes using new European codes. In Ochiai, Yasufuku & Omine (eds), *Earth Reinforcement*, Balkema, Rotterdam, pp. 943–961.

Gässler, G. & Gudehus, G. (1981). Soil nailing: Some aspects of a new technique. In Proceedings of the 10th International Conference on Soil Mechanics and Foundation Engineering, Stockholm, vol. 3, pp. 665–670.

GEO (Geotechnical Engineering Office) (2008). *Guide to Soil Nail Design and Construction (Geoguide 7)*. Geotechnical Engineering Office, Civil Engineering and Development Department, HKSAR Government, Hong Kong, 97 p.

Hirano, M. (2003). Actual status of the application of the soil nailing to expressway cut-slope construction in Japan. In Landmarks in Earth Reinforcement: Proceedings of the International Symposium on Earth Reinforcement, Japan, pp. 919–934.

JHPC (Japan Highway Public Corporation) (1987). *Guide for Design and Construction on Reinforced Slope with Steel Bars*. Japan Highway Public Corporation, Japan, 33 p.

JHPC (1998). *Design and Construction Guidelines for Soil Nailing* (in Japanese). Japan Highway Public Corporation, Japan, 111 p.

Kulczykowski, M., Przewlocki, J. & Konarzewska, B. (2017). Application of soil nailing technique for protection and preservation historical buildings. *IOP Conference Series: Materials Science and Engineering*, vol. 245, paper 022055.

Kwong, A.K.L. & Chim, J.S.S. (2015). Performance monitoring of a 50 m high 75° cut slope reinforced with soil nail bars made from glass fibre reinforced polymer (GFRP) at Ho Man Tin Station. In Proceedings of the HKIE Geotechnical Division 35th Annual Seminar, the Hong Kong Institution of Engineers, Hong Kong, pp. 141–149.

Lazarte, C.A., Robinson, H., Gomez, J.E., Baxter, A., Cadden, A. & Berg, R. (2015). Geotechnical Engineering Circular No. 7: Soil Nail Walls: Reference Manual. Federal Highway Administration, US Department of Transportation, Washington, DC, USA, Report No. FHWA-NHI-14-007, 385 p.

Mori, S. & Asahi, M. (1986). Experimental work of cut slope improved with reinforcing bars. *Civil Engineering*, vol. 41, no. 10, pp. 68–72.

Ortigao, J.A.R., Palmeira, E.M. & Zirlis, A.C. (1995). Experience with soil nailing in Brazil: 1970–1994. *Proceedings of the Institution of Civil Engineers: Geotechnical Engineering*, vol. 113, pp. 93–106.

O'Donovan, J., Feudale Foti, E., Turner, M., Irvin, C. & Mothersille, D. (2020). *Grouted Anchors and Soil Nails - Inspection, Condition Assessment and Remediation*. Construction Industry Research & Information Association, London, UK, CIRIA Report No. C794, 193 p.

Perry, J., Pedley, M. & Reid, M. (2001). Infrastructure Embankments: Condition Appraisal and Remedial Treatment. Construction Industry Research & Information Association, London, UK, CIRIA Report No. C550, 232 p.

Phear, A., Dew, C., Ozsoy, B., Wharmby, N.J., Judge, J. & Barley, A.D. (2005). Soil Nailing: Best Practice Guidance. Construction Industry Research & Information Association, London, UK, CIRIA Report No. C637, 286 p.

Powell, G. & Watkins, A. (1990). Improvement of marginally stable existing slopes by soil nailing in Hong Kong. In Proceedings of the International Reinforced Soil Conference, Glasgow, UK, pp. 241–247.

Rabcewicz, L.V. (1964a). The new Austrian tunnelling method: Part 1. *Water Power*, vol. 16, no. 11, pp. 453–457.

Rabcewicz, L.V. (1964b). The new Austrian tunnelling method: Part 2. *Water Power*, vol. 16, no. 12, pp. 511–515.

Rabcewicz, L.V. (1965). The new Austrian tunnelling method: Part 3. *Water Power*, vol. 17, no. 1, pp. 19–24.

Shen, C.K., Bang, S. & Herrman, L.R. (1981a). Ground movement analysis of an earth support system. *Journal of Geotechnical Engineering Division*, vol. 107, no. GT 12, pp. 1609–1624.

Shen, C.K., Bang, S., Romstad, K.M., Kulchin, L. & De Natale, J.S. (1981b). Field measurement of an earth support system. *Journal of Geotechnical Engineering Division*, vol. 107, no. GT 12, pp. 1625–1642.

Stocker, M.F., Korber, G.W., Gässler, G. & Gudehus, G. (1979). Soil nailing. In Proceedings of the International Conference on Soil Reinforcement: Reinforced Earth and Other Techniques, Paris, March, vol. II, pp. 469–474.

Tatsuoka, F., Kondo, K., Miki, G., Ikuhara, O., Hamada, E. & Sato, T. (1983). Fundamental study on the reinforcement of sand with tensile reinforcement. *Soils and Foundations*, vol. 31, no. 9, pp. 11–19.

Tayama, S. & Kawai, Y. (1999). The current status and outlook for soil nailing in Japan. In Special Volume for the Proceedings of the 11th Asian Regional Conference on Soil Mechanics and Geotechnical Engineering, Korea, pp. 51–61.

Chapter 2

Fundamentals and concepts

2.1 MECHANISM OF A SOIL NAILED STRUCTURE

Before carrying out a soil nail design, it is important for designers to understand the fundamental working principles underlying a soil nailed structure. In this chapter, various key factors that govern the behavior of soil nailed structures, including stability and deformation, are identified and investigated. They include the orientation and inclination of soil nails, bending and shear resistance of soil nails, nail heads and facing, as well as pullout resistance. The robustness and reliability of soil nails are also discussed using probability theory.

Soil nailing is an *in situ* ground reinforcement technique that improves the stability of earth structures, i.e. slopes, retaining walls, embankments, and excavations, principally through the mobilization of tension in the slender structural elements. In principle, soil nails act to increase the overall shear strength of the reinforced soil mass and limit the ground deformation through transferring the stresses into a more stable zone (i.e. passive zone) within the deeper part and beyond the reinforced ground mass. Unless the soil nails are specifically pre-tensioned, they are essentially passive elements and do not normally generate restoring or resisting forces until ground deformation occurs and gives rise to relative movement and frictional interaction between the ground and soil nails, hence mobilizing tensile force in the soil nails. For steep excavations or newly formed cuts, due to a reduction in lateral stresses, soil nails are in these cases progressively stressed during the course of excavation. For existing slopes, retaining walls, or embankments, soil nails normally remain unstressed until the earth structures are subject to transient loads, such as those associated with a rise in groundwater table, traffic loading, and earthquake, which can bring about destabilizing forces. As shown in Figure 2.1, the tensile forces in soil nails are primarily developed through the frictional interaction between soil nails and the ground, although there may also be some secondary effect through the reactions provided by soil nail heads and the facing. Also, the bending and shear resistances would also be generated in the process of

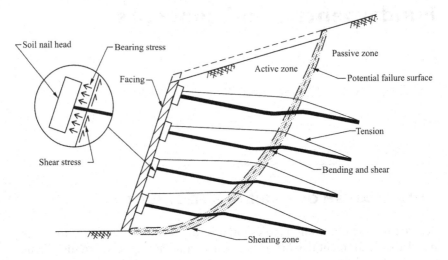

Figure 2.1 Reinforcing mechanism of a soil nailed structure.

nail-ground interaction. More discussion on the nail-ground interaction is given in Section 2.2.

Once tensile forces are developed in the soil nails, they can reinforce the soil mass by directly reducing some of the driving force and allowing a higher shearing resistance to be mobilized along the shearing surface. As illustrated in Figure 2.2, the reinforcing effect is achieved by (i) supporting some of the applied shear loading and hence reducing the average shear stress to be carried by the soil mass, and (ii) by increasing the normal stress in the soil acting on the potential failure surface, thus mobilizing a higher shear strength of the soil.

Whenever the soil mass of an earth structure is subjected to reduction of lateral pressure (e.g. by excavation) or reduction of soil strength (e.g. reduction in effective stress by inundation), there is a tendency of development of instability along a potential shearing zone. As a result, two distinct zones, namely the active zone and passive zone (or resistant zone), will be developed across the potential shear surface in the soil mass. The active zone, as its name suggests, tends to deform laterally away from the parent soil mass. The passive zone, on the other hand, is a region behind the potential shear surface where it remains intact with the parent soil mass. Soil nails can be considered as structural elements that tie the active and passive zones together. The magnitude and distribution of load developed along the soil nails depend on the response mechanism of the soil nailed structure. Where the ground deformation is small, it will be resisted and minimized essentially by the axial stiffness of the soil nails. As a soil nail is relatively stiffer than the soil, the bending stiffness of the soil nails may also be mobilized as the deformation increases, albeit of a small magnitude when compared

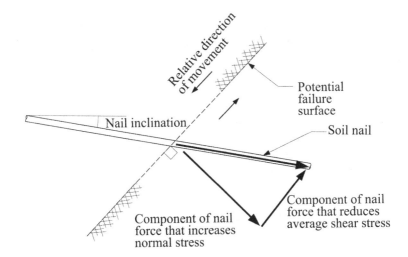

Figure 2.2 Reinforcing effect of a soil nail.

with the axial stiffness. With the increase in ground deformation, the tension in soil nails will continue to increase until it reaches the maximum stress that can be mobilized. Depending on the configuration of the earth structure and the soil nails, a deformed S-shaped soil nail within the zone of high strain (i.e. shear surface or shear zone) will be developed (Figure 2.1). Ultimate failure occurs when the soil nails rupture or when the soil moves around the soil nails.

Soil nail heads and the facing also play an important role in the reinforcing mechanism. Apart from providing a reaction for individual soil nails to mobilize tensile forces, they also provide a confinement effect by limiting the ground deformation in the direction close to normal to the soil surface. Depending on the resultant strain of the soil nail heads and facing, the load transfer mechanism involves primarily a bearing mechanism along the axis of the soil nail. It may also be a combination of bending and shearing if the strain is in a direction other than axial. The confinement effects provided by the soil nail heads and facing also help to prevent local failures near the surface of the earth structure (i.e. within the active zone), and promote integral action of the reinforced soil mass through redistribution of forces amongst the soil nails.

In practice, the internal stability of a soil nailed structure is conveniently assessed using a two-zone model, i.e. the active zone and passive zone, which are separated by a potential failure surface instead of a shearing zone (Figure 2.1). The active zone is the region in front of the potential failure surface, where it tends to detach from the soil nailed structure. The mobilization of tensile forces of soil nails in the active zone is mainly attributed to the reaction provided by the nail heads and facing, and the friction between soil nail and the surrounding ground mass. The passive zone, on the other hand, is the region behind the potential failure surface that remains more or less intact. The resistance against pullout failure of the soil nails in the passive zone is provided by the friction of soil nail that is embedded into the ground. The soil nails act to tie the active zone to the passive zone. Readers should, however, note that the two-zone configuration separated by a potential failure surface is only a simplified model to facilitate the carrying out of limit equilibrium analysis where the relative deformation of a soil nailed structure is not accounted for. In reality, in a soil slope for instance, unless the failure is dictated by joints where the failure surface is distinct, there is generally a shearing zone subject to shear distortion.

2.2 NAIL-GROUND INTERACTION

2.2.1 General

Soil nails, which act as structural elements, interact with the ground to support the stresses and strains that would otherwise cause the unreinforced ground to fail. Section 2.1 presents the overall picture about the mechanism of a soil nailed structure, and this section focuses on the details of the nail-ground interaction. Nail-ground interaction is a complex aspect in soil mechanics analysis. The forces developed in soil nails are basically through interaction between the ground, soil nails, nail heads, and facing. The forces that can be mobilized in soil nails are influenced by many factors, including the mechanical properties of soil nails (i.e. tensile strength, shear strength, and bending capacity), inclination and orientation of soil nails, shear strength of the ground, relative stiffness between soil nails and the ground, friction between soil nails and the ground, size of soil nail heads, nature of facing, etc. Although nail-ground interaction is a complicated process, there are two fundamental mechanisms that govern the interaction, namely (i) nail-ground friction that leads to the development of axial tension (or compression) in soil nails, and (ii) soil bearing stress on soil nails and the nail-ground friction on the sides of soil nails that lead to the development of shear forces and bending moments in soil nails. The development of stresses and strains in the active zone is resisted by the soil shear strength and the strength of soil nail under the combined loading of tension, bending, and shear. When there is a small ground movement in the active zone, in particular at the shearing zone where the active zone moves downward relative to the passive zone, the soil nail will experience both axial and lateral strains. The axial strain will mobilize tensile forces, and the lateral strain will mobilize shear force and bending moment in the soil nail. If the soil nail is aligned with the direction of tensile strain increment of soil, the predominant action of the soil nail will then be in tension and the corresponding shear force and bending moment induced in the soil nail will be small.

2.2.2 Orientation and inclination of soil nails

2.2.2.1 Laboratory tests

One of the advantages of using the soil nailing technique is that soil nails can be installed in the ground conveniently at various orientations or inclinations to suit the site conditions. For present purposes, orientation of the reinforcement, θ, is defined as the angle between the reinforcement and the normal to the potential shearing surface, whereas reinforcement inclination, α, is the angle between the reinforcement and the horizontal (Figure 2.3). An important design consideration is the possible effect, if any, of different orientation or inclination of soil nails on the performance of a soil nailed structure, i.e. the effectiveness in mobilizing tensile forces. The fundamental behavior of reinforced soil with respect to the orientation and inclination of reinforcement has been investigated by many researchers by means of shear box tests (e.g. Jewell, 1980; Marchal, 1986; Jewell & Wroth, 1987; Hayashi et al., 1988; Palmeira & Milligan, 1989; Davies, 2009). One of the major findings of the research work is that the increase in shear strength of the reinforced soil is dependent on the orientation of the reinforcement. The investigation shows that the introduction of reinforcement significantly modifies the state of stress and strain in the soil mass, and that by varying the orientation of the reinforcement, the soil shear strength will either be increased or decreased. Figure 2.4 compares the patterns of strain increments in the soil in unreinforced and reinforced direct shear box tests

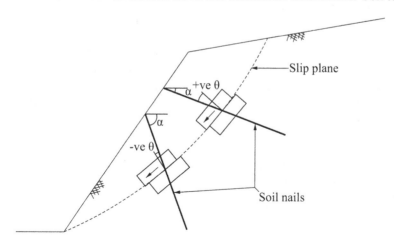

Figure 2.3 Definition of orientation, θ, and inclination, α, of nail reinforcement.

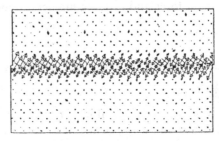

(a) Principal Strain Increments in Unreinforced Sand at Peak

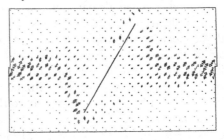

(b) Principal Strain Increments in Sand Reinforced by a Grid at Orientation $\theta = +30°$

Figure 2.4 Incremental strains at peak shearing resistance: (a) principle strain increments in unreinforced sand at peak; (b) principal strain increments in sand reinforced by a grid at orientation q = +30° (Jewell, 1980).

carried out by Jewell (1980). It is evident that the presence of reinforcement causes a significant re-orientation of the principal axes of strain increment of the soil in the vicinity of the reinforcement (see Figure 2.4b). The soil strain increments close to the reinforcement are small because the reinforcement inhibits the formation of a failure plane. As long as the reinforcement is orientated in the same direction as the tensile strain increment of soil, tensile forces will be induced in the reinforcement through the friction between the soil and the reinforcement. Likewise, compressive forces are induced in the reinforcement if the reinforcement is placed close to the compressive strain increment of the soil.

As shown in Figure 2.5, Jewell & Wroth (1987) demonstrated that tensile or compressive strain increments could be experienced by varying the orientation of the reinforcement. The shear strength of the soil starts to increase when the reinforcement is placed in the direction of tensile strain increment, and it reaches a maximum when the orientation of the reinforcement is close to the direction of the principal tensile strain increment. When the reinforcement is oriented in the direction of compressive strain increment, there is a decrease in shear strength of the reinforced soil. This suggests that in order to maximize shear strength improvement of the soil, the reinforcement should be placed in the direction of principal tensile strain increment

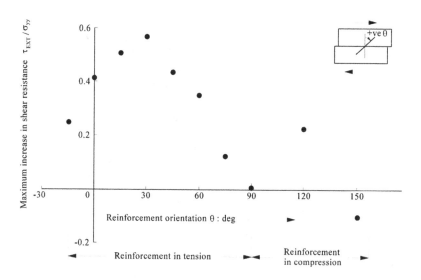

Figure 2.5 Effect of reinforcement orientation on the increase in shear strength of reinforced soil. (Modified Jewell & Wroth, 1987.)

in the soil as far as possible. When the reinforcement deviates from its optimum orientation, the strength improvement will decrease.

Jewell & Wroth (1987) further demonstrated that the tensile force in the reinforcement increased the shearing resistance by an amount, $\Delta\tau$ (Figure 2.6):

$$\Delta\tau = \frac{T}{A_s}(\cos\theta\ \tan\phi' + \sin\theta) \tag{2.1}$$

where T is the reinforcement tensile force, A_s is the area of the shearing plane in the direct shear apparatus, θ is the orientation of the reinforcement, and ϕ' is the angle of shearing resistance of the soil.

The above equation indicates that the reinforcing effect of the reinforcement on the soil shear strength depends on the mobilized tensile force, T, and the orientation of the reinforcement, θ. It may be noted from Figure 2.5 that the reinforcing effect (i.e. the increase in soil shear strength) reaches a maximum when the reinforcement orientation is close to about 30°, which is likely the direction of the principal tensile strain increment of the soil. However, when the reinforcement orientation increases, the reinforcing

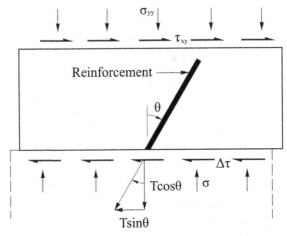

Legend:

Δτ   Increase of shearing resistance of soil

A_s Area of soil shearing plane

T   Reinforcement tensile force

Figure 2.6 Increase of shearing strength of soil due to tensile force in reinforcement. (Modified Jewell & Wroth, 1987.)

effect diminishes substantially until at a point where the effect becomes negative, i.e. reinforcement is in compression.

2.2.2.2 Numerical simulations

Apart from laboratory tests on individual reinforcement, Shiu & Chang (2006) carried out numerical simulations using the finite difference method coupled with the strength reduction technique to examine the effects of soil nail inclination on the reinforcing action in a 20 m high model slope at an angle of 55°. The initial shear strength parameters of the soil were assumed to be: effective cohesion c' = 10 kPa and angle of shearing resistance ϕ' = 43°. The factor of safety (FS) of the slope before applying soil nails was found to be 1.0. Seven rows of soil nails with a spacing of 2.5 m (vertical) and 1.5 m (horizontal) were provided. By varying the inclination of the soil nails from 0° to 65° below the horizontal, the corresponding tensile forces mobilized in the soil nails and the increase in the factor of safety, ΔFS, would vary (see the results summarized in Table 2.1). The increase in the factor of safety is close to 1.0 (i.e. 100%) with little variations for the range of α between 0° and 20°. The ΔFS decreases substantially as α further increases beyond 20°, reflecting that the reinforcing action of the nails reduces rapidly with increasing nail inclinations. When α equals 65°,

Table 2.1 Effect of soil nail inclination, α, on the mobilization of tensile forces in soil nails

Inclination of Soil Nails, α (Below Horizontal)	Total Tensile Forces Mobilized in Soil Nails (kN/m)	Increase in Factor of Safety, ΔFS
0°	949	1.0 (100%)
5°	933	1.0 (100%)
10°	981	1.0 (100%)
20°	974	1.0 (100%)
30°	850	0.85 (85%)
45°	725	0.5 (50%)
50°	342	0.2 (20%)
55°	60	0.15 (5%)
65°	0	0 (0%)

the ΔFS is close to 0. This indicates that steeply inclined soil nails do not provide much stabilizing effect to the slope.

As an illustration, Figure 2.7 compares the distribution of axial forces mobilized in soil nails inclined at an angle of 20° and 55°, respectively. One may notice that when $\alpha = 20°$, all the axial forces in soil nails are in tension (with a total magnitude of 974 kN/m, Table 2.1), whereas those in the upper four rows of soil nails become compression when $\alpha = 55°$ and the corresponding orientations, θ, are greater than 90° (with the total magnitude of tensile force reducing to 60 kN/m). In other words, compressive forces rather than tensile forces could be developed in steeply inclined soil nails, which would be detrimental to the reinforcing effect on the slope.

2.2.2.3 Full-scale field tests

In 2004, Guler & Bozkurt (2004) conducted a full-scale field test in which two walls with soil nails inclined above and below the horizontal were constructed. Strain gauges and survey points were mounted on the soil nails and the walls respectively to monitor their performance. A major observation is that the tension that was mobilized in the soil nails inclined above horizontal was some 10% to 15% greater than these below horizontal. Furthermore, the deformation of the wall with upward inclined nails was less.

The results of laboratory tests, numerical simulations, and field trials as described above indicate that the angle at which the soil nails are installed has a significant effect on the performance of the soil nailed structure. From a theoretical point of view, the effectiveness of soil nails in terms of mobilization of tension reaches a maximum when their orientation is close to the direction of the principal tensile strain increment of the soil.

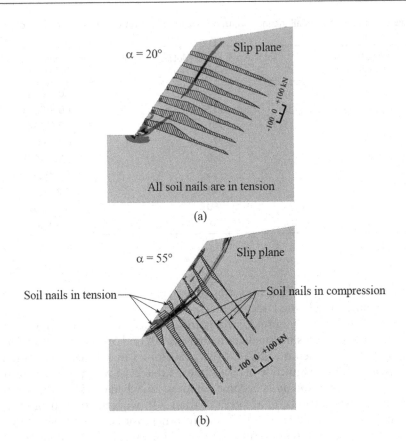

Figure 2.7 Distribution of axial forces in soil nails with inclination of α = 20° and α = 55° (Shiu & Chang, 2006).

Figure 2.8 shows schematically the more effective orientation of soil nails in mobilizing tensile forces with respect to the potential failure slip. The optimal inclination of the soil nails varies from above horizontal near the slope crest to below horizontal when they are close to the slope toe. For ease of construction, soil nails are seldom installed at an inclination above the horizontal in particular for those installed using the drill-and-grout method. Also, Johnson et al. (2002) suggested that as the inclination of soil nails below the horizontal is increased, the length of soil nails in the passive (resistant) zone will be increased (Figure 2.9). This increase in length in the passive zone accounts for the decrease in the effectiveness due to the angle of inclination. The extra overburden above the soil nails in this case also increases the pullout resistance (see Chapter 6 for more details about the pullout resistance of soil nails). Nonetheless, the effectiveness of soil nails in overall terms decreases as their inclination below the horizontal increases.

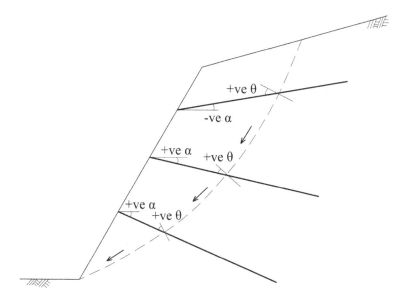

Figure 2.8 Optimum orientation of soil nails.

In summary, if the soil nails are aligned close to the direction of the principal tensile strain increment of the soil, the forces mobilized in the soil nails will be primarily in tension, which is developed through the mechanism of nail-ground friction. In contrast, if the soil nails are aligned in the direction of compressive strain increment in the soil, compressive forces will be developed in the soil nails. This can lead to a decrease in normal stresses in the soil on the potential failure surface, which reduces the shearing resistance of the reinforced soil mass. The above principles explain the effect of the soil nail orientation and inclination on the mobilization of tensile forces in soil nails. In general, the effectiveness of a soil nail in mobilizing tensile force decreases as the inclination of the soil nail to the horizontal, α, increases (see Figure 2.10). For most soils, given that the soil nails are mostly sub-horizontally inclined (about 5° to 20° below horizontal), the minimum deformation required to mobilize the full bending and shear resistance of a soil nail is about one order of magnitude greater than that required to mobilize the full tensile force, and hence the primary action of soil nails in service is in tension (Clouterre, 1991; Byrne et al., 1998). The effects of bending and shear resistance of soil nails will be further discussed in Section 2.2.3. If the soil nails are steeply inclined, their effectiveness will be reduced significantly as some of the soil nails may be in compression. This will reduce the stability of a soil nailed structure. Therefore, steeply inclined soil nails should be used with caution and their effects should be properly accounted for.

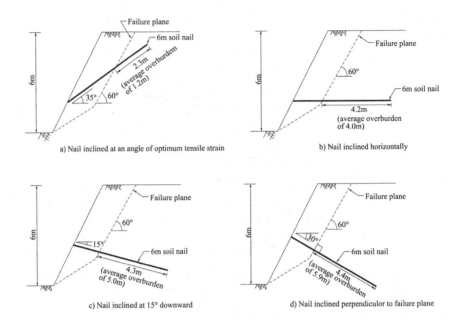

Figure 2.9 Relationship between soil nail inclination and overburden and length of soil nail in resistant zone (Johnson et al., 2002).

2.2.3 Bending and shear resistance of soil nails

Although soil nails work predominantly in tension, they can also sustain shear forces and bending moments, and such ability may also enhance the shear strength of soil. The development of shear force in soil nails involves a mechanism which is dependent on the relative stiffness of soil nail and ground mass, soil bearing strength, orientation and shear deformation of reinforcement, and shear zone thickness. In general, the shear and bending resistances of soil nails are mobilized only after relatively large displacement has taken place along the shearing zone of the soil mass. Some researchers have found that the shear and bending resistances of soil nails contribute no more than 3% of the overall stability of a soil nailed structure (e.g. Pedley, 1990).

The ability of soil nails to increase the shear strength of soil by acting under combined loading of shear and tension is one of the more controversial aspects of design. The presence of axial force in reinforcement reduces the maximum bending moment that can be sustained. In turn, the maximum shear force that can be developed in the reinforcement depends on the maximum bending moment. This means there is a connection between the maximum axial force and the maximum shear force that a soil nail can support. The effect of bending and shear resistance of soil nails has been investigated by many researchers (e.g. Schlosser, 1982; Marchal, 1986; Gigan &

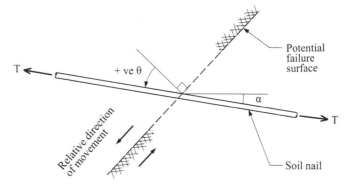

(a) Mobilization of Tensile Force in a Soil Nail

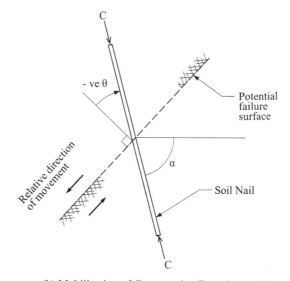

(b) Mobilization of Compressive Force in a Soil Nail

Legend:

α   Inclination of soil nail to the horizontal

θ   Orientation of soil nail with respect to the potential failure surface

Figure 2.10 Effect of soil nail inclination on the mobilization of force in a soil nail.

Delmas, 1987; Plumelle et al., 1990; Pedley, 1990; Jewell & Pedley, 1990 & 1992; Davies et al., 1993; Bridle & Davies, 1997; Davies & Le Masurier, 1997; Smith & Su, 1997; Tan et al., 2000; Cheung & Chang, 2012). When a soil nail is being sheared, lateral soil pressures will be developed on both sides of the soil nail, and hence the development of axial forces (tension or compression), bending moment, and shear force. Figure 2.11 reproduces some experimental results which show the distribution of bending moment, shear force, and lateral soil stress in shear box tests reported by Pedley (1990). The results are presented in the form of the distributions of bending moment (M), shear force (P_s), and lateral stress on the reinforcement

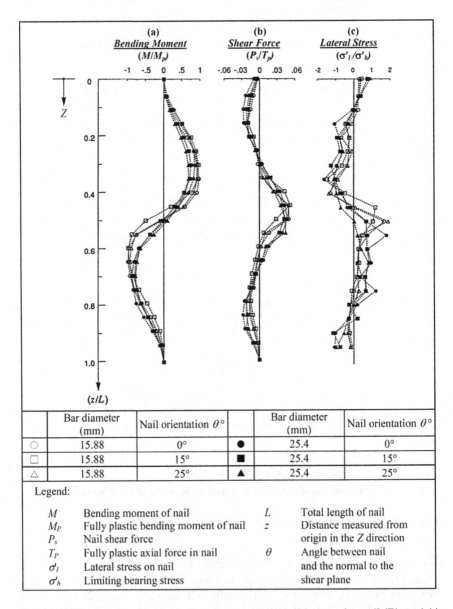

	Bar diameter (mm)	Nail orientation $\theta °$		Bar diameter (mm)	Nail orientation $\theta °$
○	15.88	0°	●	25.4	0°
□	15.88	15°	■	25.4	15°
△	15.88	25°	▲	25.4	25°

Legend:

M	Bending moment of nail	L	Total length of nail
M_P	Fully plastic bending moment of nail	z	Distance measured from origin in the Z direction
P_s	Nail shear force		
T_P	Fully plastic axial force in nail	θ	Angle between nail and the normal to the shear plane
σ_l	Lateral stress on nail		
σ_b	Limiting bearing stress		

Figure 2.11 Distribution of: (a) bending moment (M/M_p); (b) shear force (P_s/T_p); and (c) lateral stress (σ_l/σ_b) at a shear displacement of 60 mm (Pedley, 1990).

(σ_l), normalized by the plastic moment capacity (M_p), plastic axial capacity (T_p), and the limiting soil bearing stress (σ_b), respectively. The distributions were the stress conditions of the reinforcement at a soil shear displacement of 60 mm. One may reveal that even when the bending moments were close to the fully plastic values (i.e. $M/M_p = 1$), the maximum shear forces (P_s) in the reinforcement that could be mobilized were less than 6% of the plastic axial capacity (T_p). It is envisaged that the reinforcement shear force has a much smaller influence on soil shear strength improvement than the reinforcement axial force for two reasons: (i) a component of the reinforcement shear force reduces the normal stress on the slip plane and thereby decreases the soil strength; and (ii) the limiting reinforcement shear force is only a small percentage of the axial capacity of the reinforcement. This shows that reinforcement acting in shear and bending is much less efficient than that in axial tension.

Numerical simulations of direct shear box and zone shear box tests have been carried out by Cheung & Chang (2012) to examine reinforcement-soil interaction in respect of mobilization of axial force, shear force, and bending moment in the reinforcement, as well as the effect of shear zone thickness under direct shearing. The simulated shear box has dimensions of 1 m wide, 3 m deep, and a length of 6 m split into two halves. A 1 m wide steel plate equivalent to 40 nos. of 25 mm diameter and 3 m long steel bars was installed in the middle of the box across the vertical slip surface (Figure 2.12). The box was filled with homogeneous sand. Shear displacements of 5 mm, 10 mm, 25 mm, and 50 mm were imposed. Figures 2.13 and 2.14 show the shear forces and bending moments along the reinforcement at a shear displacement of 5 mm. The results indicate that the mobilization of axial force, shear force, and bending moment in the reinforcement for a given shear displacement depends on the width of the shear zone. The narrower the shear zone, the higher the shear force and bending moment will be. The degree of mobilization of axial force of the reinforcement for a given shear displacement is negligible, whereas that of bending moment

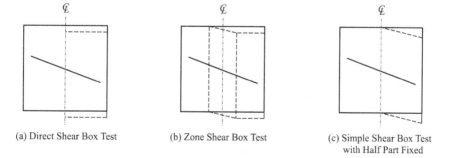

(a) Direct Shear Box Test   (b) Zone Shear Box Test   (c) Simple Shear Box Test with Half Part Fixed

Figure 2.12 Cases considered in numerical study of effect of shear zone (Cheung & Chang, 2012).

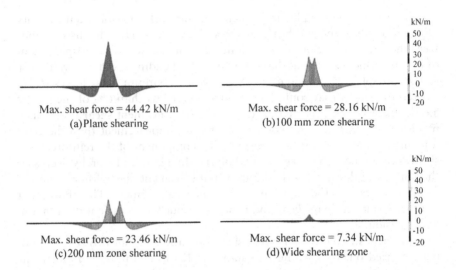

Figure 2.13 Shear force along reinforcement in shear box model at shear displacement of 5 mm (Cheung & Chang, 2012).

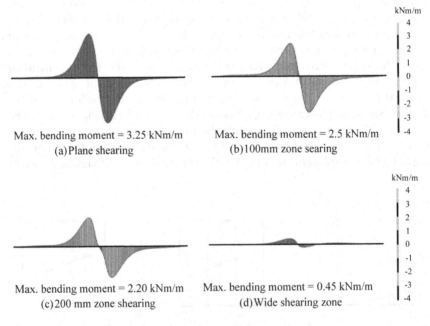

Figure 2.14 Bending moment along reinforcement in shear box model at shear displacement of 5 mm (Cheung & Chang, 2012).

is the highest among the three actions, i.e. axial force, shear force, and bending moment. For example, at a plan shear displacement of 5 mm, the maximum shear force mobilized was found to be about 0.5% of the ultimate tensile capacity of the reinforcement (Figure 2.13), whereas the corresponding maximum bending moment was about 7% of the plastic moment (Figure 2.14). Indeed, when the shear displacement increased to 50 mm, the maximum shear force to the tensile capacity and bending moment to the plastic moment became 2.5% and 37%, respectively. In other words, it is likely that bending failure will occur first if the reinforcement is subject to direct shearing. The results of the numerical simulations generally agree with those determined using laboratory tests (e.g. Pedley, 1990). In addition, the sharper the contrast soil stiffness between the two shearing zones, the higher the mobilization of axial force, shear force, and bending moment will be in the reinforcement for a given shear displacement.

Although there has been some debate amongst researchers in the past as to the behavior of soil reinforcement under combined loadings, there appears to be a common understanding that the reinforcing action of soil nails is predominantly derived from the axial tensile forces in nails, with the beneficial effect of shear and bending capacity of nail being of secondary importance only. Furthermore, it is commonly accepted that the shear forces and bending moments are mobilized in soil nails only when a nailed-structure has large deformations and is close to failure. Pedley (1990) back-analyzed the monitoring results of an instrumented 6 m high soil nailed wall that was loaded to failure. He found that the largest contribution of reinforcement shear force to soil strength improvement was less than 3% of that due to reinforcement axial force. This is in agreement with the theoretical study results that only a small amount of shear force can be mobilized in soil nails.

Nowadays, it is widely accepted that the contributing effect of shear and bending resistance to the stability of soil nailed structures is small as compared with the tensile resistance. However, due to its ductile behavior, the shear resistance of steel reinforcement does provide beneficial effect when the soil nailed structure is close to failure. The mobilization of shear and bending resistance at large displacement will result in a more ductile failure mode which is more desirable than the undesirable brittle failure mode. As a result, the design will err on the safe side, while the shear and bending resistance is ignored. Although steel bars are commonly used for nail reinforcement, reinforcement of other types of materials may be considered. Some materials such as fiber reinforced polymer may exhibit brittle behavior at failure. In this case, the performance of these materials in response to the combined actions of tension, shear, and bending should be assessed critically during design and construction stages. More information about this subject can be found in Section 4.10 of Chapter 4 and Section 5.7 of Chapter 5.

2.2.4 Soil nail heads and facing

The main functions of soil nail heads and facing are to improve the tension mobilization efficiency in soil nail reinforcement, enhance local ground stability, as well as limit the ground deformation in the active zone. The functions that can be provided by soil nail heads and facing depend on the stiffness and the surface coverage of the nail heads and facing, and the stiffness and shear strength of the soil underneath. Although many studies have been carried out on soil nailing in the past decades, our understanding of the role of soil nail heads and facing on the behavioral mechanisms of a soil nailed structure is not as good as that of soil nail reinforcement (Byrne et al., 1998; Sanvitale et al., 2013). This is partly due to the lack of good quality field monitoring data. The available data from instrumented soil nails are generally difficult to interpret in the vicinity of soil nail heads and facing, where the bending effect of soil nail tends to be more significant as a result of the weight of the soil nail head (Thompson & Miller, 1990). There have been little field monitoring data obtained using load cells at soil nail heads probably because of the difficulties in placing load cells between soil nail heads and soil (Stocker & Reidinger, 1990).

Despite the lack of good quality field monitoring data, a number of studies including numerical simulations (e.g. Ehrlich et al., 1996; Shiu & Chang, 2005; Cheung & Chan, 2010), model tests (e.g. Gutierrez & Tatsuoka, 1988; Tei et al., 1998; Sanvitale et al., 2013), and full-scale field tests (e.g. Gässler & Gudehus, 1981; Plumelle & Schlosser, 1990; Muramatsu et al., 1992) have been carried out to better understand the role of soil nail heads and facing in the mechanism of a soil nailed structure. Some of the studies are related to soil nailed retaining walls where the nail heads are integrated into the concrete facing. In summary, these studies suggest that soil nail heads and facing can have a significant effect on the performance of a soil nailed structure. The presence of soil nail heads and facing enables the reinforced ground mass to sustain a higher load before failure compared to that without soil nail heads or facing. In addition, the load that a reinforced structure can sustain increases as the stiffness of the soil nail heads and facing increases. The soil nail heads and facing also help to limit the amount of ground deformation, in particular when the reinforced structure is subject to seepage conditions. The following sections endeavor better understanding of the role of soil nail heads and facing in the behavior of a nailed structure through numerical simulations, model tests, and full-scale field trials.

2.2.4.1 Numerical simulations

In Hong Kong, soil nail heads used in slope stabilization works are usually in the form of isolated reinforced concrete pads. To investigate the effect of soil nail heads on the stability of soil nailed slopes, a series of numerical simulations using the two-dimensional finite difference code, Fast

Nail parameters

Grout hole diameter = 100 mm
Bar diameter = 40 mm
Soil nail length = 20 m
Vertical spacing = 2.5 m
Horizontal spacing = 1.5 m
f_y = 460 N/mm^2

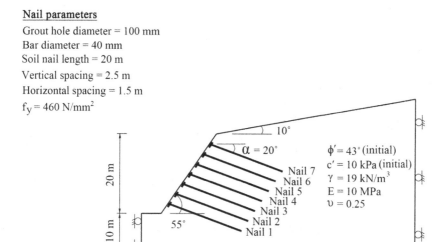

Figure 2.15 Configuration and parameters of the model slope (Shiu & Chang, 2005).

Lagrangian Analysis of Continua (FLAC), was conducted on a model slope of 20 m high at an angle of 55° as shown in Figure 2.15 (Shiu & Chang, 2005). In the simulations, soil nail heads of various sizes commonly used in local practice (from 0.4 m × 0.4 m × 0.25 m thick, 0.8 m × 0.8 m × 0.25 m thick to full coverage of slope surface) were considered. For illustration, Figure 2.16 shows one of the analyses that compares the tensile forces mobilized in the soil nails without heads with those in nails with heads of 0.8 m wide. The strength reduction technique was used to examine the effects of nail head size on the reinforcing action in the reinforced model slope. The initial soil shear strength parameters of effective cohesion c' = 10 kPa and angle of shearing resistance ϕ' = 43° were adopted. The factor of safety (*FS*) of the unreinforced model slope was found to be 1.0, whereas the *FS*s corresponding to the two reinforced cases (without and with nail heads) were

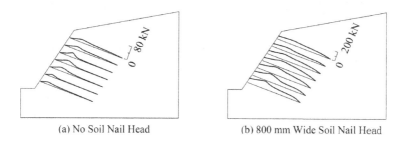

(a) No Soil Nail Head   (b) 800 mm Wide Soil Nail Head

Figure 2.16 Distribution of tensile forces along soil nails with and without nail heads (Shiu & Chang, 2005).

found to be 1.2 and 2.2, respectively. It is clearly seen that no tensile force is developed at the front end of the soil nails for the case without nail heads, whereas for the case with heads, large tensile forces are mobilized in the soil nails at the connections to the nail heads. One may also notice that the maximum tensile forces mobilized along the soil nails are much larger in the latter case. The simulations suggest that the presence of soil nail heads can significantly enhance the tension mobilization efficiency of the soil nail reinforcement, and hence the stability of the soil nailed slope.

Shiu & Chang (2005) further investigated the relationship between the size of nail head and the reinforcing effect on the model slope, with the results summarized in Table 2.2. It is evident that the degree of tensile force mobilization in the soil nails increased as the size of nail head increased. However, the forces mobilized in the soil nails were not directly proportional to the size of nail heads and there was a diminishing return when the size of nail head was kept increasing. And the diminishing return is obvious when the size of nail head was increased from 0.8 m × 0.8 m to full coverage on the model slope surface (i.e. 2.5 m × 1.5 m). Apart from this, Shiu & Chang (2005) found from the analyses that the ratio of the tensile forces developed at the nail heads to their corresponding maximum tensile forces varies from 0.55 to 0.75. This suggests that the presence of nail heads enhances the degree of tensile force mobilization of the soil nails, and hence the performance of a soil nailed structure.

Cheung & Chan (2010) carried out a separate numerical study aimed at investigating the confinement effect of soil nails heads on ground deformation. A numerical model was established using the finite element code developed by PLAXIS BV, which is applicable to two-dimensional analysis of deformations and stability in geotechnical engineering. The mechanical behavior of the soil was modeled as linear-elastic, perfectly-plastic, Mohr-Coulomb (MC) model. As illustrated by the results of numerical analysis shown in Figure 2.17, the mean effective stress and hence the shearing resistance of the soil behind the 0.4 m wide nail heads is larger than that without nail heads. This suggests that the shear strength of soil, and hence

Table 2.2 Relationship between size of nail head and the mobilization of tensile forces in soil nails

Size of Nail Head	Total of Maximum Tensile Forces Mobilized in Soil Nails (kN/m)	Factor of Safety (FS)	Increase in Factor of Safety, ΔFS
Unreinforced	–	1.0	–
No nail head	247	1.2	0.2 (20%)
0.4 m × 0.4 m	957	2.0	1.0 (100%)
0.8 m × 0.8 m	1,211	2.2	1.2 (120%)
Full coverage (2.5 m × 1.5 m)	1,220	2.4	1.4 (140%)

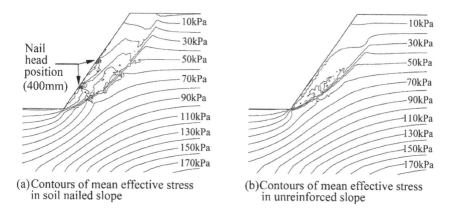

(a) Contours of mean effective stress in soil nailed slope

(b) Contours of mean effective stress in unreinforced slope

Figure 2.17 Contours of mean effective stress in the model soil nailed slope and the unreinforced slope (Cheung & Chan, 2010).

the stability of the slope, is improved by the presence of soil nail heads as a result of the corresponding confinement effect. This also helps to reduce the deformation of the reinforced soil mass. However, Cheung & Chan (2010) also pointed out that the forces developed in the soil nails were not directly proportional to the size of nail heads and there was a diminishing return as the size of nail heads was kept increasing, which is consistent with the findings by Shiu & Chang (2005).

2.2.4.2 Model tests

A series of centrifuge model tests has been conducted by the Hong Kong University of Science and Technology to investigate the reinforcing effect of soil nails, nail heads, and grillage beams (Ng et al., 2007; Zhou, 2008). The tests were carried out at 30 g on model slopes of completely decomposed granitic compacted fill with an angle of 65° and an equivalent height of 15 m. The model slopes included an unreinforced slope, nailed slopes with different sizes of nail heads, and nailed slopes with grillage beams. For the nailed slopes, the nails were inclined at an angle of 20° below horizontal. Porewater pressures, slope deformation, axial nail forces, and contact pressures between soil and nail heads were measured. Figure 2.18 shows an instrumented model used in one of the nailed slope centrifuge tests.

At 30 g, vertical tension cracks and a continuous failure surface were observed at the crest and within the soil mass of the unreinforced slope, respectively. In contrast, when the nailed slopes were at 30 g, the reinforced soil masses, either with nail heads/grillage beams or not, were found intact. However, the soil deformation at the crest of the nailed slopes with nail heads or grillage beams was found smaller than that of the nailed slopes without them. Furthermore, the deformation was found decreasing with an

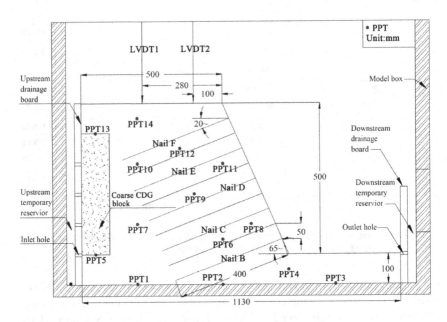

Figure 2.18 Set-up of a soil nailed slope model in centrifuge test.

increase of the surface coverage of nail heads or grillage beams. The study indicates that the presence of soil nail heads or grillage beams enhanced the mobilization of tension in the soil nails and reduced the soil deformation. For the model without soil nail heads or grillage beams, the maximum tensile force was developed approximately at middle length of the reinforcement, but the location of the maximum tensile force moved toward the slope surface when soil nail heads or grillage beams were provided. In addition, the maximum mobilized tensile forces were found increasing with an increase of the size of nail heads. The test results support and agree with the findings of the numerical simulations that presence and size of soil nail heads or grillage beams can noticeably enhance the mobilization of tensile forces in the soil nails and improve the stability and deformation control of a soil nailed slope.

Apart from centrifuge tests, Gutierrez & Tatsuoka (1988) performed loading tests on three model sand slopes, viz. (i) unreinforced slope, (ii) slope reinforced with metal strips but without a facing, and (iii) slope reinforced with metal strips and with a facing. The slopes were loaded at the crest by a footing with a smooth base (Figure 2.19). The results of the tests, as shown in Figure 2.20, indicate that the reinforced slope with facing can sustain a higher load (as represented by the average normal stress between the footing and the slope crest) than the case of reinforced slope without facing, and a much higher load than that in the unreinforced slope. Figure 2.21 depicts the corresponding shear strain contours of the loading tests.

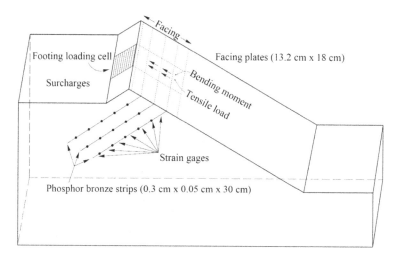

Figure 2.19 Model sand slope reinforced with metal strips and with a facing (Gutierrez & Tatsuoka, 1988).

One may observe that deep and shallow failure planes developed in the unreinforced slope. For the reinforced slope without facing, failure took place close to the slope face, whereas for the reinforced slope with a facing, failure is observed at a greater depth. This is because, in the case of without a facing, the tensile capacity of the reinforcement may not be fully utilized and thus renders the system relatively ineffective in retaining the active zone. When a facing was provided, it enhanced the stability of the reinforced slope and helped prevent shallow failures.

Separate model tests were conducted by Sanvitale et al. (2013) in which the focus was on the influence of the axial and flexural stiffness of the facing on the reinforcing effect. Table 2.3 summarizes the physical characteristics of the facings used in the tests.

The results of the investigation indicate that the stiffness of the facing could greatly influence the mobilization and distribution of tensile forces along the soil nails. For example, Figure 2.22 shows that the mobilization of tensile force in the case of a high stiffness facing is much higher than that of a low stiffness facing. This means the reinforced slopes with high stiffness facing are capable of withstanding a higher load than those with less stiff facing. In addition, the locus of the maximum tensile force moves toward the facing (i.e. the active zone is reduced), as the stiffness of the facing increases. In other words, the distribution of tensile force in the active zone in the case of a high stiffness facing is largely uniform as compared with that in the case of a low stiffness facing. This suggests that a high stiffness facing has reduced the relative soil nail displacement in the active zone. The coverage of the facing also affected the mobilization of tensile

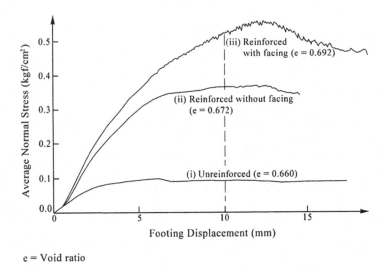

e = Void ratio

Figure 2.20 Strengthening effect of reinforcement and facing (Gutierrez & Tatsuoka, 1988).

forces in the soil nails. It was revealed that a 100% coverage of facing could mobilize much higher tensile forces than the same type of facing, with only 25% coverage (see the results of facing PMMA (polymethyl methacrylate) and PMMA25 in Figure 2.22).

2.2.4.3 Full-scale field tests

Although limited, some full-scale field tests were carried out by researchers in the past decades (e.g. Gässler & Gudehus, 1981; Plumelle & Schlosser, 1990; Muramatsu et al., 1992). For example, Muramatsu et al. (1992) conducted a full-scale field trial where soil nails of 5 m length at a spacing of 1.5 m, both vertically and horizontally, were installed by the direct driving method during excavation works. The final cutting was at a gradient of 80° and 9.5 m high. The geology of the site mainly comprised talus type deposits consisting of primarily gravel and boulders. In the trial, two types of facing, namely 0.1 m thick sprayed concrete and concrete crib (0.2 m wide at spacing of 1.5 m), were tested.

The horizontal displacement and tension were measured by inclinometers and strain gauges mounted along the soil nails respectively at different stages of excavation. In general, the displacement increased as the excavation proceeded. The maximum displacements measured near the cut faces with the two types of facing are found comparable. However, the displacement along the soil nails with a crib facing was largely uniform but for the case of sprayed concrete facing, displacement decreased with the increasing

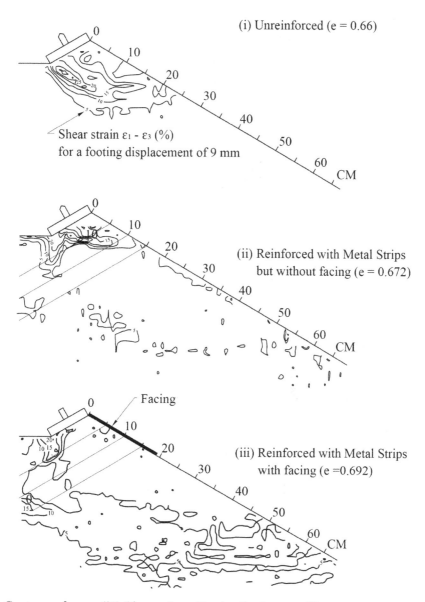

(i) Unreinforced (e = 0.66)

Shear strain $\varepsilon_1 - \varepsilon_3$ (%)
for a footing displacement of 9 mm

(ii) Reinforced with Metal Strips
but without facing (e = 0.672)

(iii) Reinforced with Metal Strips
with facing (e = 0.692)

Facing

Contours of $\varepsilon_1 - \varepsilon_3$ (%) Observed at a Footing Settlement of 9 mm
e = Void ratio

Figure 2.21 Shear strain contour (Gutierrez & Tatsuoka, 1988).

Table 2.3 Physical characteristics of facings in the model tests (Modified Sanvitale et al., 2013)

Type of Facing	Coverage (%)	Thickness/Wire Dia. (mm)	Wire Spacing (mm)	Young Modulus, E (GPa)	Axial Stiffness, EA/m (N/mm)	Flexural Stiffness, EI/m (Nmm²/mm)
PMMA	100	4	–	3.2	12,800	17,066.67
Steel Mesh	100	1 (Dia.)	6	210	26,180	3,318.06
Brass	100	0.25	–	126	31,500	236.25
Steel Net	100	0.24 (Dia.)	1.02	70	3,105	22.66
PMMA95	95	4	–	3.2	–	–
PMMA25	25	4	–	3.2	–	–

Note: PMMA and PMMA95 are polymethyl methacrylate facings with 100% and 95% coverage respectively.

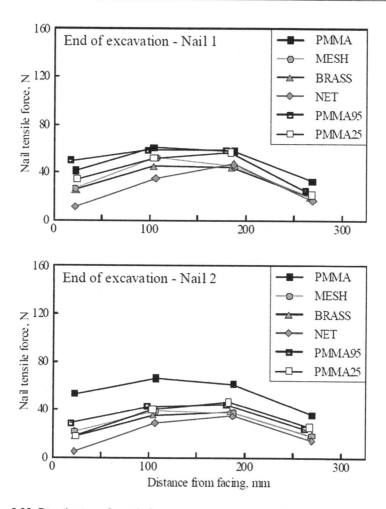

Figure 2.22 Distribution of tensile forces along upper nail (nail 1) and lower nail (nail 2) with facing of different stiffness (Sanvitale et al., 2013).

distance from the slope surface. This suggests that the presence of the crib facing has enhanced the integral action of the reinforced zone. In contrast to the case of sprayed concrete facing where the maximum tensile force occurs near the middle of soil nails, the maximum tensile force occurred near the surface if concrete crib facing is adopted. This is in line with the findings from the numerical and model tests that the locus of maximum tensile force in nails usually moves toward the facing if its stiffness increases. In addition, the tensile force that could be mobilized by the presence of crib facing is higher than the case of sprayed concrete facing. This suggests that a concrete crib facing with a higher stiffness is able to enhance the overall slope stability by a larger margin than that with a sprayed concrete facing.

In summary, the observations from numerical simulations, model tests, and field trials highlight the importance of the presence of soil nail heads and facing, in particular a facing of high rigidity, in enhancing the stability and deformation control of a soil nailed structure. Nail heads or facing with high stiffness improve the mobilization efficiency of soil nail forces, and hence the overall stability and deformation control of a reinforced structure.

2.2.5 Pullout resistance of soil nails

Soil nails improve the stability of an earth structure principally through the mobilization of tension in the nails. Whenever there is relative movement between the ground and the soil nail, shear stress will be mobilized at the nail-ground interface. Indeed, the stability of the entire reinforced structure is supported by the resistance against pullout failure of the part of soil nail that is embedded into the ground behind the potential failure surface (i.e. passive zone). As with other soil-structure interaction problems, pullout failure is governed by soil strength and stress state. Although a number of research projects on pullout resistance have been carried out, the mechanism of the bonding between soil nails and the surrounding soil is still not thoroughly understood to date. This is largely due to the complexity of the soil nail installation and stress mobilization process. Many studies, including numerical simulations, laboratory tests, and field tests, indicate that the pullout resistance of a soil nail is affected by numerous factors including nail type, installation method, soil type, soil shear strength, surface roughness of the nail, grouting pressure, soil stress state, groundwater condition, etc. More details on the study of pullout resistance are given in Chapter 6.

2.3 DEFORMATION OF SOIL NAILED STRUCTURES

Soil nails are basically passive elements. The nail forces can only be mobilized whenever there is ground deformation, either during or after construction, which leads to frictional interaction between the ground and soil nails. Nailed structures such as slopes, retaining structures, embankments, and excavations will deform during the progressive mobilization of tensile forces in the soil nails; it is essential to understand the deformation behavior of these reinforced structures in order to assess the effect on the structures, utilities, services, etc., within the influence zone of the deformed structures. In theory, a soil nailed structure should be flexible enough to allow the structure to deform, so as to mobilize tension in the soil nails. At the same time, the structure should be sufficiently rigid to permit load sharing among the soil nails. Field monitoring data indicate that deformation may continue after the end of construction. Deformation is also sensitive to climatic effects such as freezing and thawing. Thus, the soil nailing technique should not be used for those structures that cannot tolerate considerable movement during their service life. The amount of deformation that a soil nailed structure would experience depends on a number of factors, *inter alia*, the geological and hydrogeological setting of the site, physical characteristics of the soil mass (e.g. stiffness), configuration of soil nails such as spacing, orientation, inclination, length, installation method, presence and types of nail heads or facing, etc.

The authors have collected information from the literature on the performance of soil nailed excavations in different countries. Table 2.4 summarizes the details of the soil nailed excavations at 24 construction sites spreading over Hong Kong, Iran, Ireland, Japan, Turkey, and the United States where soil nails have been installed to provide lateral support with the deformation monitored during the course of excavation (Thompson & Miller, 1990; Shiu et al., 1997; Tayama & Kawai, 1999; Armour & Cotton, 2003; Durgunoglu et al., 2007; Menkiti & Long, 2008; Kwong & Chim, 2015; Cetin, 2016; Zolqadr et al., 2016). The excavation depth varies from 7 m to 50 m, and the gradient of the excavation face varies mostly from about 60° to sub-vertical.

For the present discussion, the performance of the soil nailed excavation is reflected by the maximum horizontal displacement, $\delta_{h,max}$, measured at the top of excavation to the corresponding maximum excavation depth, D_{max}. Among the 56 field measurements at 24 construction sites, the normalized maximum horizontal displacement (i.e. $\delta_{h, max}/D_{max}$) is found to vary from negligible to 0.5%, with most of them (53 out of 56) being less than 0.4% (Figure 2.23). More importantly, no anomaly or adverse effects on the structures, utilities, or services in the vicinity of these construction sites were reported. According to Clouterre (1991), horizontal displacement, δ_h, is related to the mobilization of tensile force in the soil nails, whereas vertical displacement, δ_v, tends to relate to the bending moment in the soil nails. The field measurements at these sites suggest that soil nails have been effective in reinforcing the excavations where the maximum normalized horizontal displacements at the top of excavations were generally within the acceptable range of below 0.4%, as proposed by Clouterre (1991).

As an illustration, Figure 2.24 shows the profile of horizontal ground displacement during a 7 m deep excavation with 100 mm thick shotcrete facing at the trial site in Kochi, Japan (Tayama & Kawai, 1999). Five rows of soil nails were installed in the course of excavation. One may notice that the magnitude of the horizontal ground displacement increased, in particular near the crest, as the depth of excavation increased. Also, the maximum horizontal ground displacement occurred at about 1 m to 2 m below the crest. The corresponding mobilization of nail forces during the excavation and progressive installation of soil nails are shown in Figure 2.25. As indicated in Figure 2.25, tensile forces were mobilized progressively from top to bottom layers of soil nails. Meanwhile, the locus of maximum tensile forces in the soil nails moved away from the excavation surface as the excavation proceeded. The actual behavior will also be affected by the type and

Table 2.4 Summary of soil nailed excavations at 24 construction sites in Hong Kong, Iran, Ireland, Japan, Turkey, and the United States with deformation monitored during excavation

Site	Geology	Max. Excavation Depth, D_{max}	Gradient of Excavation	Details of Reinforcement						Max. Displacement at Top of Excavation		Reference
				Material	Diameter	Inclination	Length	Spacing	Slope Facing	Horizontal, $\delta_{h, max}$	Normalized Horizontal, $\delta_{h, max}/D_{max}$	
Hong Kong	Fill, completely and highly decomposed granite	26 m, 24 m (50 m total)	45°, 75°	Glass fiber reinforced polymer	40 mm	20°	20 m	2 m (v) × 2 m (h) in fill 1.3 m (v) × 2 m (h) for others	100 mm shotcrete	70 mm	0.14%	Kwong & Chim (2015)
Hong Kong	Completely decomposed granite	13.5 m	80°	Steel	32 mm and 40 mm	10°	10–11 m	1.5 m (v) × 1 m (h)	100 mm shotcrete	13 mm	0.1%	Shiu et al. (1997)
Tehran, Iran	Gravel/ sand with some clay and silt	29.3 m	90°	Steel	32 mm	10–15°	6–14 m	1–2.5 m	Shotcrete	35 mm	0.12%	Zolqadr et al. (2016)
Dublin, Ireland	Glacial till	12 m	80°	Steel	20 mm, 25 mm & 32 mm	10°	8 m, 10 m, & 11 m	1.5–1.9 m	75–110 mm	16 mm	0.13%	Menkiti & Long (2008)

(Continued)

Table 2.4 (Continued) Summary of soil nailed excavations at 24 construction sites in Hong Kong, Iran, Ireland, Japan, Turkey, and the United States with deformation monitored during excavation

Site	Geology	Max. Excavation Depth, D_{max}	Gradient of Excavation	Details of Reinforcement						Max. Displacement at Top of Excavation		Reference
				Material	Diameter	Inclination	Length	Spacing	Slope Facing	Horizontal, $\delta_{h,max}$	Normalized Horizontal, $\delta_{h,max}/D_{max}$	
Kochi, Japan	Gravel/weathered slate	3.5 m 3.5 m (7.0 m total)	1:0.3 (~73°) 1:0.0 (~90°)	Steel	32 mm	–	5 m	1.0 m × 1.0	100 mm shotcrete	9.58 mm	0.14%	Tayama & Kawai (1999)
Chitose, Japan	Talus/weathered sandstone	2.0 m 1.1 m 9.9 m (13.0 m total)	1:0.5 (~63°) 1:0.5 (~90°) 1:0.2 (~79°)	Steel	22 mm 19 mm	–	4.0 m 3.0 m	1.2 m × 1.2 m 1.5 m × 1.5 m	50/100 mm shotcrete/sprayed frame	2.60 mm	0.02%	
Oita, Japan	Talus/sandstone/conglomerate	8.0 m (total)	1:0.5 (~63°)	Steel	25 mm	–	4 m	1.0 m × 1.0 m	100 mm shotcrete	2.51 mm	0.03%	
Shirotori, Japan	Welded tuff	9.3 m 7.3 m (16.6 m total)	1:0.5 (~63°) 1:0.0 (~90°)	Steel	19 mm	–	2 m	1.5 m × 1.5 m	50/150 mm shotcrete	6.00 mm	0.04%	

(Continued)

Table 2.4 (Continued) Summary of soil nailed excavations at 24 construction sites in Hong Kong, Iran, Ireland, Japan, Turkey, and the United States with deformation monitored during excavation

Site	Geology	Max. Excavation Depth, D_{max}	Gradient of Excavation	Details of Reinforcement						Max. Displacement at Top of Excavation		Reference
				Material	Diameter	Inclination	Length	Spacing	Slope Facing	Horizontal, $\delta_{h,max}$	Normalized Horizontal, $\delta_{h,max}/D_{max}$	
Kiyomi, Japan	Talus/granodiorite porphyry	5.8 m 2.7 m 4.0 m (12.5 m total)	1:0.3 (~73°) 1:0.0 (~90°)	Steel	25 mm	–	4.5 m 2.0 m	1.2 m × 1.2 m 1.5 m × 1.5 m	50/150 mm shotcrete	4.58 mm	0.04%	
Ohzu, Japan	Weathered schist/ fractured zones	12.8 m	1:0.5 (~63°)	Steel	32 mm	–	6 m	1.2 m × 1.2 m	50 mm shotcrete/ sprayed frame	0.50 mm	0.00%	
Besiktas, Istanbul, Turkey	Extensively fractured sandstone	18.5 m	85°	Steel	–	10°	9.2 m	1.5 m (v) × 1.8 m (h)	Shotcrete	14.8 mm	0.08%	Durgunoglu et al. (2007), Cetin (2016)
		25.0 m 32.5 m					11.2 m 10.2 m			22.5 mm 52 mm	0.09% 0.16%	

(Continued)

Table 2.4 (Continued) Summary of soil nailed excavations at 24 construction sites in Hong Kong, Iran, Ireland, Japan, Turkey, and the United States with deformation monitored during excavation

Site	Geology	Max. Excavation Depth, D_{max}	Gradient of Excavation	Details of Reinforcement						Max. Displacement at Top of Excavation		Reference
				Material	Diameter	Inclination	Length	Spacing	Slope Facing	Horizontal, $\delta_{h,max}$	Normalized Horizontal, $\delta_{h,max}/D_{max}$	
Istinye Sariyer, Istanbul, Turkey	Extensively fractured siltstone, claystone	10 m	80°	Steel	—	10°	5.1 m	1.5 m (v) × 2.0 m (h)	Shotcrete	27 mm	0.27%	
		10 m					8.3 m			22 mm	0.22%	
		12 m					6.0 m			37.2 mm	0.31%	
		12 m					8.8 m			25.2 mm	0.21%	
		14 m					9.1 m			19.6 mm	0.14%	
		16 m					8.2 m			44.8 mm	0.28%	
		18 m					9.3 m			55.8 mm	0.31%	
		20 m					9.7 m			80 mm	0.4%	
		22 m					10.1 m			96.8 mm	0.44%	
Levent, Istanbul, Turkey	Extensively fractured sandstone, siltstone, claystone	14 m	85°	Steel	—	10°	8.4 m	1.5 m (v) × 2.0 m (h)	Shotcrete	28 mm	0.2%	
		15.7 m					9.4 m			45.5 mm	0.29%	
		18.8 m					9.5 m			31.9 mm	0.17%	
		21.3 m					11.6 m			68.2 mm	0.32%	
		25.3 m					11.2 m			58.2 mm	0.23%	

(Continued)

Table 2.4 (Continued) Summary of soil nailed excavations at 24 construction sites in Hong Kong, Iran, Ireland, Japan, Turkey, and the United States with deformation monitored during excavation

Site	Geology	Max. Excavation Depth, D_{max}	Gradient of Excavation	Details of Reinforcement					Slope Facing	Max. Displacement at Top of Excavation		Reference
				Material	Diameter	Inclination	Length	Spacing		Horizontal, $\delta_{h,max}$	Normalized Horizontal, $\delta_{h,max}/D_{max}$	
		26.3 m					11.8 m			86.8 mm	0.33%	
		28.3 m					11.3 m			53.7 mm	0.19%	
		28.3 m					11.3 m			67.9 mm	0.24%	
		28.3 m					11.6 m			96.2 mm	0.34%	
Maslak, Istanbul, Turkey	Extensively fractured siltstone, claystone	18.3 m	85°	Steel	—	10°	6.7 m	1.5 m (v) ×1.6 m (h)	Shotcrete	58.6 mm	0.32%	
Maltepe, Istanbul, Turkey	Extensively fractured sandstone, siltstone, claystone	7 m	85°	Steel	—	15°	6.4 m	1.5 m (v) × 1.8 m (h)	Shotcrete	5.6 mm	0.08%	
		9 m					12.0 m			16.2 mm	0.18%	
		9 m					7.3 m			15.3 mm	0.17%	
		10 m					12.0 m			24.0 mm	0.24%	
Pendik, Istanbul, Turkey	Extensively fractured sandstone	14.7 m	85°	Steel	—	10°	9.3 m	2.0 m (v) × 2.0 m (h)	Shotcrete	10.3 mm	0.07%	
		16.2 m					9.6 m	2.0 m (v) × 1.8 m (h)		17.8 mm	0.11%	
		18.4 m								12.9 mm	0.07%	

(Continued)

Table 2.4 (Continued) Summary of soil nailed excavations at 24 construction sites in Hong Kong, Iran, Ireland, Japan, Turkey, and the United States with deformation monitored during excavation

Site	Geology	Max. Excavation Depth, D_{max}	Gradient of Excavation	Details of Reinforcement						Max. Displacement at Top of Excavation		Reference
				Material	Diameter	Inclination	Length	Spacing	Slope Facing	Horizontal, $\delta_{h,max}$	Normalized Horizontal, $\delta_{h,max}/D_{max}$	
Sisli, Istanbul, Turkey	Fractured sandstone	22 m	—	Steel	—	—	—	1.5 m (v) × 1.5 m (h)	Shotcrete	14.6 mm	0.07%	
Levent, Istanbul, Turkey	Highly fractured graywacke	25 m	—	Steel	—	—	—	1.5 m (v) × 2.0 m (h)	Shotcrete	45 mm	0.18%	
		24 m								14.4 mm	0.06%	
		20 m								10 mm	0.05%	
		19.5 m								9.75 mm	0.05%	
		24 m								19.2 mm	0.08%	
Levent, Istanbul, Turkey	Dense fill	9.5 m	—	Steel	—	—	—	1.5 m (v) × 2.0 m (h)	Shotcrete	7.6 mm	0.08%	
		10 m								10 mm	0.1%	
		10 m								10 mm	0.1%	
		13 m								22.1 mm	0.17%	
Umraniye, Istanbul, Turkey	Fractured sandstone	15.2 m	—	Steel	—	—	—	1.6 m (v) × 1.4 m (h)	Shotcrete	24.9 mm	0.16%	
		15.2 m								34.5 mm	0.23%	

(Continued)

Table 2.4 (Continued) Summary of soil nailed excavations at 24 construction sites in Hong Kong, Iran, Ireland, Japan, Turkey, and the United States with deformation monitored during excavation

| Site | Geology | Max. Excavation Depth, D_{max} | Gradient of Excavation | Details of Reinforcement | | | | | | Max. Displacement at Top of Excavation | | Reference |
				Material	Diameter	Inclination	Length	Spacing	Slope Facing	Horizontal, $\delta_{h, max}$	Normalized Horizontal, $\delta_{h, max}/D_{max}$	
Kucuk-cek-mece, Istanbul, Turkey	Sandstone, siltstone, claystone	7.5 m 7.5 m	–	Steel	–	–	–	2.0 m (v) × 2.0 m (h)	Shotcrete	15.4 mm 38.4 mm	0.20% 0.51%	Armour & Cotton (2003)
Seattle, Wash-ington, US	Fill (2–3 m), very dense sandy silt with some gravel	14 m	~90°	Steel	–	15°	5–12 m	1.8 m (v) × 1.8 m (h)	100 mm shotcrete	< 6.4 mm	< 0.05%	
Bellevue, Wash-ington, US	Fill (1–4 m), dense sandy silt with little gravel	26 m	~90°	Steel	25 mm, 32 mm	15°	9–20 m	1.8 m (v) × 1.8 m (h)	300 mm shotcrete	< 6.4 mm	< 0.02%	

(Continued)

Table 2.4 (Continued) Summary of soil nailed excavations at 24 construction sites in Hong Kong, Iran, Ireland, Japan, Turkey, and the United States with deformation monitored during excavation

| Site | Geology | Max. Excavation Depth, D_{max} | Gradient of Excavation | Details of Reinforcement | | | | | | Max. Displacement at Top of Excavation | | | Reference |
				Material	Diameter	Inclination	Length	Spacing	Slope Facing	Horizontal, $\delta_{h, max}$	Normalized Horizontal, $\delta_{h, max}/D_{max}$		
Seattle, Washington, US	Fill (2–4 m), very dense glacial outwash sand and gravel, very dense lacustrine fine sand & silt	16.8 m	~90°	Steel	25 mm, 32 mm	20° (top row) 15°	10.7 m	1.8 m (v) × 1.8 m (h)	300 mm shotcrete	18 mm	0.1%	Thompson & Miller (1990)

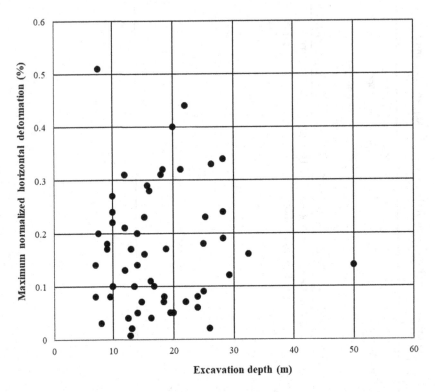

Figure 2.23 Maximum normalized horizontal deformation vs excavation depth.

stiffness of the slope facing. One would expect that if the facing is flexible, the maximum tensile force would occur near the middle of the reinforcement, whereas it would be closer to the surface if the stiffness of the facing is increased. For the present case, the distributions of nail forces, as shown in Figure 2.25, suggest that the shotcrete facing is relatively flexible. In general, a rigid facing can better limit the ground deformation and improve overall stability by enhancing the tension mobilization efficiency of the soil nails. Nonetheless, the maximum horizontal ground displacement at the top of excavation at this site (9.58 mm) was measured to be about 0.14% of the maximum excavation depth (7.0 m), which is considered acceptable.

2.4 ROBUSTNESS AND RELIABILITY OF SOIL NAILS

2.4.1 Robustness

The design of a structure is normally oriented toward normal circumstances which can be anticipated to exist during the service life of the structure. Limiting the design to this, however, will leave the structure vulnerable

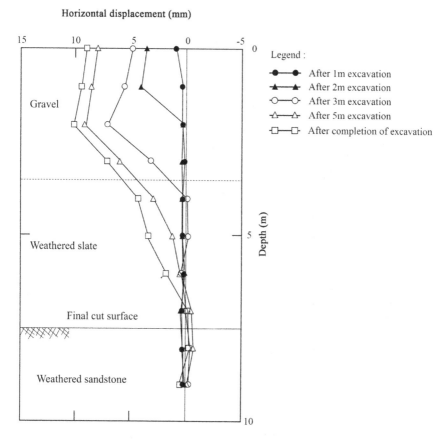

Figure 2.24 Ground displacement profile during excavation at the site in Kochi, Japan. (Modified Tayama & Kawai, 1999.)

to the events or conditions that are not included in the set of anticipated circumstances. Thus, provisions should be given to the structures during design stage to resist a set of foreseeable unforeseen conditions, such as the possible adverse impacts from climate change and seismicity. Robustness is the property of a structure that enables it to survive under unforeseen or inconceivable adverse conditions. In order to achieve this, the structure should have the ability to accommodate without failure events or actions that were not foreseen or explicitly included in the design. In addition, the response of the structure should also be insensitive to variations of the known parameters within their anticipated range of uncertainty.

In the context of soil nailing, robustness is the ability of a soil nailed structure to withstand events or conditions, such as unforeseen unfavorable ground and groundwater conditions, impacts and consequences of human error, without being damaged to an extent disproportionate to the causes.

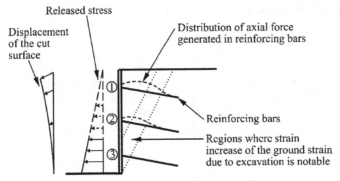

Released stress

Displacement
of the cut
surface

Distribution of axial force
generated in reinforcing bars

Reinforcing bars

Regions where strain
increase of the ground strain
due to excavation is notable

(1) After excavation of the third layer

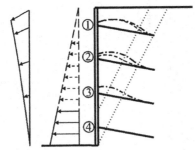

(2) After excavation of the fourth layer

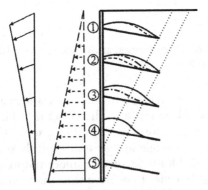

- - - - Distribution of the axial
force after excavation of
the third layer

- - - Distribution of the axial
force after excavation of
the fourth layer

——— Distribution of the axial
force after excavation of
the fifth layer

(3) After excavation of the fifth layer

Figure 2.25 Mobilization of tensile force in soil nails during excavation at the site in
Kochi, Japan (Tayama & Kawai, 1999).

Robustness of a soil nailed structure should be distinguished from "resilience," which refers to the ability of the structure to withstand shocks and stress, and be recovered after it has suffered damage. The soil nailing technique involves installation of a large number of slender structural elements into an earth structure to provide integral action with the reinforced ground mass. Because of this, it tends to have a higher system redundancy than other conventional earth retaining techniques, such as prestressed ground anchors. Thus, a soil nailed structure would be considered robust as all its members, i.e. soil nails, are resilient so that local failures can be repaired without failing the complete structure. Good track records also demonstrate that soil nailed structures can perform well under seismic loading conditions. Furthermore, it is likely that a soil nailed structure will display deformations before failure, and as such the failure tends to be ductile.

2.4.2 Reliability

In engineering problems, neither load nor resistance can be determined exactly no matter how extensive the site exploration, soil testing, and sophisticated the calculation models are. In other words, both load and resistance can be treated as variables and probability theory can be invoked in the analysis. For the sake of illustration, Figure 2.26 shows schematically the probability density functions of load, L, and resistance, R, where a probability density function is a probability distribution showing the relative likelihood of the occurrence of the value of the continuous random variable. If we define the factor of safety of a soil nailed structure, FS, to be the ratio of resistance to load, R/L, one would expect a variation of factor of safety as depicted, in the form of a probability density function, as in Figure 2.27.

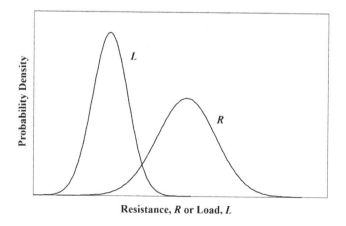

Figure 2.26 Probability density functions of resistance, R, and load, L.

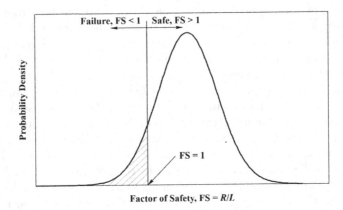

Figure 2.27 Probability density function of factor of safety, FS.

Furthermore, if "failure" or "non-performance" is defined as an event with the factor of safety, FS, being less than unity, the probability of failure or non-performance, P_f, can be expressed as:

$$P_f = P(FS < 1) \tag{2.2}$$

or

$$P_f = P(G < 0) \tag{2.3}$$

where G is the performance function and is defined as FS−1

Graphically, Equation (2.2) is the area under the probability density curve of factor of safety less than unity as hatched in Figure 2.27. Thus, the reliability of a nailed structure can be represented by the probability of performing its intended function satisfactorily, which is herein the event with the factor of safety being greater than or equal to unity. In practice, we seldom have sufficient information on the full probability density functions. Instead, it is more practicable to determine statistical descriptors (e.g. mean and covariances) of the pertinent variables such that the reliability of a nailed structure can be evaluated based on these descriptors. As illustrated in Figure 2.28, the distance between the mean value of the factor of safety, μ_{FS}, and that at the limiting boundary, FS = 1, can be expressed as the standard deviation of FS, σ_{FS}, multiplied by a quantity called reliability index, β. As the reliability index, β, increases, one would expect that the probability of failure (hatched area) decreases (i.e. the reliability of the nailed structure increases). Therefore, the reliability of a nailed structure can be simply indicated by this index.

During the course of developing the reliability index concept in the early 1970s, a problem was realized that the index depended on the way

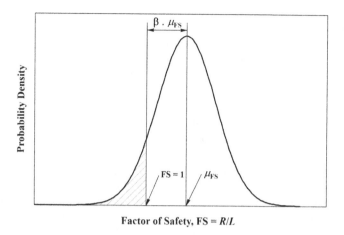

Figure 2.28 Graphical representation of reliability index, β.

of formulating the performance function. That is, the reliability index corresponding to the performance function of $G = R/L-1$ is different from that for $G = R-L$. This problem was subsequently resolved by Hasofer & Lind (1974) by using the first order Taylor series expansion about the most probable failure point and the first two statistical descriptors (i.e. mean and covariance) to express the reliability index, β. This index that represents the reliability of an engineering system was found to be invariant to the way of formulating the performance function.

In the context of soil nailed structures, the load, L, and the resistance, R, are functions of multiple variables, X, such as nail forces, soil shear strength parameters, piezometric pressure response, and imposed loading. If the performance function of a nailed structure is $G(X)$, then $G(X) = 0$ becomes the limiting hyper-surface separating the "safe" domain, i.e. $G(X) > 0$ from the "failure" or "non-performance" domain, i.e. $G(X) < 0$. In other words, the performance of a nailed structure can be represented by the performance function:

$$G(X) = G(X_1, X_2, \ldots X_n) \tag{2.4}$$

where $X = (X_1, X_2, \ldots, X_n)$ is a vector of the pertinent design random variables such as nail forces, soil shear strength parameters, piezometric pressure response, and imposed loading.

In general, for n number of design random variables, the reliability index, β, can be expressed in the following matrix form (Ditlevsen, 1981):

$$\beta = \min_{x \in F} \sqrt{(x - \mu)^T C^{-1} (x - \mu)} \tag{2.5}$$

where x = vector of design variables, μ = vector of mean values, and C = covariance matrix.

For illustrative purposes, one may consider that the soil shear strength parameters (i.e. effective cohesion, c' and angle of shearing resistance, ϕ') are the only two pertinent design random variables with means and standard deviations of μ_c', μ_Φ', and σ_c', σ_Φ' respectively. The limiting surface (i.e. $G(X) = 0$) is then reduced to a line $L–L$ as shown in Figure 2.29. Recall Equation (2.5) above, the reliability index, β, can be visualized as the minimization of an ellipsoidal quadratic form subject to a constraint that the ellipsoid just touches the limiting surface. Graphically, a reliability index, β, is expressed as the ratio between the ellipse tangent to the limiting surface $L–L$ and that with one standard deviation. Once the statistical descriptors of the relevant design random variables are identified and quantified, the corresponding reliability index, β, can be evaluated by simple calculations using the appropriate stability model, such as the method of slices coupled with the limit equilibrium method for a nailed slope. Upon obtaining the reliability index, β, the corresponding failure probability can be obtained by $\Phi(-\beta)$, where $\Phi()$ is the cumulative normal distribution function.

In 2002, the Geotechnical Engineering Office in Hong Kong carried out a study to investigate the reliability of soil nailed slopes (GEO, 2002). In this study, reliability analysis was carried out to compare the safety performance of a steep slope with soil nails installed and that of an unreinforced slope cut back to a gentler angle. Both slopes have originated from the same typical previously cut slope in saprolite with slope height = 20 m and slope

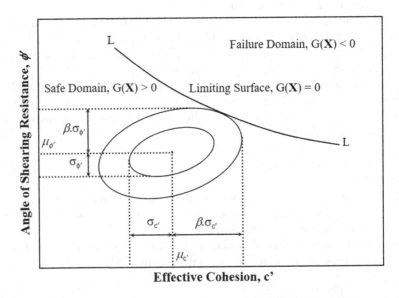

Figure 2.29 Graphical representation of reliability index, β, in nailed structure problem.

angle = 55°. They were designed to the same safety standard using the conventional factor of safety approach (i.e. both the nailed steep slope and the unreinforced gentle slope were designed to have the same factor of safety, $FS = 1.4$).

For the slope reinforced by soil nails, eight rows of soil nails were installed at a spacing of 2 m horizontally, whereas for the unreinforced slope, the slope angle was reduced to 40°. Although both the soil nailed slope and the trimmed slope have the same calculated factor of safety of 1.4, the results of reliability analyses indicate that their reliability levels are significantly different. The nailed slope has a probability of failure of about 1 in 500,000 (reliability index, $\beta =$ 4.6), whereas that for the unreinforced trimmed slope was about 1 in 1,000 ($\beta =$ 3.0). In other words, the soil nailed slope is significantly more reliable than the unreinforced albeit less steep slope. This suggests that the conventional design standards, which are based largely on deterministic approaches and are commonly adopted in routine practice, may not be able to consider such redundancy effect. In contrast, such effect could be accounted for in a risk assessment in evaluating the probability of failure (Phear et al., 2005).

One should note that the approach of assessing the reliability of a slope as described above is only associated with the critical slip surface with the lowest reliability index or the highest probability of failure. There is a growing trend of assessing the system reliability of a slope, i.e. the total probability of failure associated with many potential slip surfaces. It is anticipated that this total probability of failure of the slope is greater than that associated with any particular potential slip. Readers are recommended to refer to other publications for more details about system reliability of slopes (e.g. Zhang et al., 2011; Metya et al., 2017).

2.5 CONCLUDING REMARKS

Soil nailing is an *in situ* ground reinforcement technique that improves the stability of earth structures by transferring loads imposed on them to the ground principally through the mobilization of tension in the soil nails. Soil nails are basically passive, unstressed structural elements upon installation where tensile forces are generated only when ground deformation occurs giving rise to relative movement and frictional interaction between the ground and the soil nails. The tensile forces in soil nails are primarily developed through the frictional interaction between the soil nails and the ground, with reactions provided by soil nail heads and facing as well. In this chapter, the mechanism of a soil nailed earth structure and nail-ground interaction, including the effect of orientation and inclination of soil nails, soil nail heads and facing, pullout resistance, and bending and shear resistance of soil nails on the performance of a nailed structure, have been examined. In general, the effectiveness of a soil nail in mobilizing tensile force decreases as the inclination of the soil nail to the horizontal increases beyond about 20°. Soil nail heads and facing, in particular those with high stiffness, can enhance the stability and deformation control of a soil nailed structure. Also, the contribution of bending and shear resistance by the soil nails in the process of nail-ground interaction is relatively insignificant as the soil displacement required to mobilize such resistance is much greater than that required for tensile forces.

As the soil nailing technique involves installation of a large number of reinforcing elements into the ground to promote integral action with the soil mass, it has a higher system redundancy than other conventional earth retaining techniques. It is likely that failure of individual soil nails will result in excessive deformation and the failure is expected to be ductile. Reliability analysis also indicates that a soil nailed earth structure is more reliable than that stabilized by other conventional means, such as cutting back scheme to a gentle slope angle. The former has a much lower calculated probability of failure than the latter, even though they have the same calculated factor of safety based on limit equilibrium analysis. The good track records of soil nailing also demonstrate that many soil nailed structures have performed well without excessive displacement.

REFERENCES

Armour, T.A. & Cotton, D.M. (2003). Recent advances in soil nailed earth retention. In Earth Retention Systems: A Joint Conference. ASCE Metropolitan Section Geotechnical Group.

Bridle, R.J. & Davies, M.C.R. (1997). Analysis of soil nailing using tension and shear: Experimental observations and assessment. *Proceedings of the Institution of Civil Engineers: Geotechnical Engineering*, vol. 125, no. 3, pp. 155–167.

Byrne, R.J., Cotton, D., Porterfield, J., Wolschlag, C. & Ueblacker, G. (1998). Manual for Design and Construction Monitoring of Soil Nail Walls. Federal Highway Administration, US Department of Transportation, Washington, DC, USA, Report No. FHWA-SA-96-069R, 530 p.

Cetin, C. (2016). Performance of soil-nailed and anchored walls based on field monitoring data in different soil conditions in Istanbul. *Acta Geotechnica Slovenica*, vol. 13, no. 1, pp. 48–63.

Cheung, S.P.Y. & Chan, C.H.W. (2010). Study on Nail Head-Ground Interaction using Numerical Modelling. Geotechnical Engineering Office, Civil Engineering and Development Department, HKSAR Government, Hong Kong, GEO Report No. 251, 133 p.

Cheung, W.M. & Chang, G.W.K. (2012). Study of Nail-Soil Interaction by Numerical Shear Box Test Simulations. Geotechnical Engineering Office, Civil Engineering and Development Department, HKSAR Government, Hong Kong, GEO Report No. 265, 42 p.

Clouterre (1991). French National Research Project Clouterre: Recommendations Clouterre (English Translation 1993). Federal Highway Administration, US Department of Transportation, Washington, D.C., USA, Report No. FHWA-SA-93-026, 321 p.

Davies, M.C.R. (2009). The performance of soil nailed systems. In *Ground Improvement Technologies and Case Histories*. Geotechnical Society of Singapore, pp. 107–121.

Davies, M.C.R. & Le Masurier, J.W. (1997). Soil/Nail interaction mechanism from large direct shear tests. In Proceedings of the 3rd International Conference on Ground Improvement GeoSystems, London, pp. 493–499.

Davies, M.C.R., Jacobs, C.D. & Bridle, R.J. (1993). An experimental investigation of soil nailing. In Retaining Structures, Proceedings of the Conference of Retaining Structures, Thomas Telford, July 20–23.

Ditlevsen, O. (1981). *Uncertainty Modeling with Applications to Multi-dimensional Civil Engineering Systems*. McGraw-Hill Book Co., Inc., New York.

Durgunoglu, H.T., Keskin, H.B., Kulac, H.F., Ikiz, S. & Karadayilar, T. (2007). Performance of very deep temporary soil nailed walls in Istanbul. In TC17-Ground Improvement Workshop, WG G "Earth Reinforcement in Cut", XIV ISSMGE, Madrid, pp. 1–16.

Ehrlich, M., Almeida, M.S.S. & Lima, A.M. (1996). Parametric numerical analyses of soil nailing systems. In Proceedings Symposium on Earth Reinforcement, Balkema, Rotterdam, Fukuoka, Japan, pp. 747–752.

Gässler, G. & Gudehus, G. (1981). Soil nailing: Some aspects of a new technique. In Proceedings of 10th International Conference on Soil Mechanics and Foundation Engineering, Session 12, Stockholm, vol. 3, pp. 665–670

GEO (Geotechnical Engineering Office) (2002). *Investigation of Robust Design Method for Soil Nails: Soil Nail Reliability (Halcrow China Limited and Ove Arup & Partners Hong Kong Limited.)*. Geotechnical Engineering Office, Civil Engineering and Development Department, HKSAR Government, Hong Kong, 24 p.

Gigan, J.P. & Delmas, P. (1987). Mobilisation of Stresses in Nailed Structures (English translation). Transport and Road Research Laboratory, Contractor Report 25.

Guler, E. & Bozkurt, C.F. (2004). The effect of upward nail inclination to the stability of soil nailed structures. In Geotechnical Special Publication. ASCE, Reston, VA, vol. 126, pp. 2213–2220.

Gutierrez V. & Tatsuoka, F. (1988). Roles of facing in reinforcing cohesionless soil slopes by means of metal strips. In Proceedings of the International Geotechnical Symposium on Theory and Practice of Earth Reinforcement, IS Kyushu '88, Balkema, Rotterdam, Fukuoka, pp. 289–294.

Hasofer, A.M. & Lind, N. (1974). An exact and invariant first-order reliability format. *Journal of Engineering Mechanics*, vol. 100, no. 1, pp. 111–121.

Hayashi, S., Ochiai, H., Yoshimoto, A., Sato, K. & Kitamura, T. (1988). Functions and effects of reinforcing materials in earth reinforcement. In Proceedings of International Geotechnical Symposium on Theory and Practice of Earth Reinforcement, Fukuoka, 5–7 October, pp. 99–104.

Jewell, R.A. (1980). Some Effects of Reinforcement on the Mechanical Behaviour of Soils. PhD thesis, University of Cambridge.

Jewell, R.A. & Pedley, M.J. (1990). Soil nailing design: The role of bending stiffness. *Ground Engineering*, vol. 23, no. 2, pp. 30–36.

Jewell, R.A. & Pedley, M.J. (1992). Analysis for soil reinforcement with bending stiffness. *Journal of Geotechnical Engineering*, vol. 118, no. 10, pp. 1505–1528.

Jewell, R.A. & Wroth, C.P. (1987). Direct shear tests on reinforced sand. *Géotechnique*, vol. 37, no. 1, pp. 53–68.

Johnson, P.E., Card, G.B. & Darley, P. (2002). Soil Nailing for Slopes. TRL Limited, TRL Report No. 537, 54 p.

Kwong, A.K.L. & Chim, J.S.S. (2015). Performance monitoring of a 50 m high 75° cut slope reinforced with soil nail bars made from glass fibre reinforced polymer (GFRP) at Ho Man Tin Station. In Proceedings of the HKIE Geotechnical Division 35th Annual Seminar, the Hong Kong Institution of Engineers, Hong Kong, pp. 141–149.

Marchal, J. (1986). Soil Nail: Experimental Laboratory Study of Soil Nail Interaction (English Translation). Transport and Road Research Laboratory, Department of Transport, Contractor Report No. 239.

Menkiti, C.O. & Long, M. (2008). Performance of soil nails in Dublin glacial till. *Canadian Geotechnical Journal*, vol. 45, pp. 1685–1698.

Metya, S., Mukhopadhyay, T., Adhikari, S. & Bhattacharya, G. (2017). System reliability analysis of soil slopes with general slip surfaces using multivariate adaptive regression splines. *Computers and Geotechnics* 87, pp. 212–228.

Muramatsu, M., Nagura, K., Sueoka, T., Suami, K. & Kitamura, T. (1992). Stability analysis for reinforced cut slopes with facing. In Proceedings of the International Symposium on Earth Reinforcement Practice (edited by H. Ochiai, S. Hayashi & J. Otani), Fukuoka, Kyushu, Japan, 11–12 November, pp. 503–508.

Ng, C.W.W., Pun, W.K., Kwok, S.S.K., Cheuk, C.Y. & Lee, D.D.M. (2007). Centrifuge modelling in engineering practice in Hong Kong. Geotechnical Advancements in Hong Kong since 1970s. In Proceedings of the HKIE Geotechnical Division Annual Seminar, Hong Kong, pp. 55–68.

Palmeira, M. & Milligan, G.W.E. (1989). Large scale direct shear tests on reinforced soil. *Soils and Foundations*, vol. 29, no. 1, pp. 18–30.

Pedley, M.J. (1990). The Performance of Soil Reinforcement in Bending and Shear. PhD thesis, University of Oxford.

Phear, A., Dew, C., Ozsoy, B., Wharmby, N.J., Judge, J. & Barley, A.D. (2005). Soil Nailing: Best Practice Guidance. Construction Industry Research & Information Association, London, UK, CIRIA Report No. C637, 286 p.

Plumelle, C. & Schlosser, F. (1990). A French national research project on soil nailing: Clouterre. In *Performance of Reinforced Soil Structures*, British Geotechnical Society (edited by A. McGown, K.C. Yeo & K.Z. Andrawes), pp. 219–223.

Plumelle, C., Schlosser, F., Delage, P., & Knochenmus, G. (1990). French national research project on soil nailing. In Proceedings of the Conference of Design and Performance of Earth Retaining Structures, ASCE Geotechnical Special Publication No. 25, 18–21 June, pp. 660–675.

Sanvitale, N., Simonini, P., Bisson, A. & Cola, S. (2013). Role of the facing on the behaviour of soil-nailed slopes under surcharge loading. In Proceedings of the 18th International Conference on Soil Mechanics and Geotechnical Engineering (ICSMGE), Paris, vol. 3, pp. 2091–2094.

Schlosser, F. (1982). Behaviour and design of soil nailing. In Proceedings of Symposium on Recent Developments in Ground Improvements, Bangkok, 29 Nov.–3 Dec., pp. 399–413.

Shiu, Y.K. & Chang, G.W.K. (2005). Soil Nail Head Review. Geotechnical Engineering Office, Civil Engineering and Development Department, HKSAR Government, Hong Kong, GEO Report No. 175, 106 p.

Shiu, Y.K. & Chang, G.W.K. (2006). Effect of Inclination, Length Pattern and Bending Stiffness of Soil Nails on Behaviour of Nailed Structures. Geotechnical Engineering Office, Civil Engineering and Development Department, HKSAR Government, Hong Kong, GEO Report No. 197, 116 p.

Shiu, Y.K., Yung, P.C.Y. & Wong, C.K. (1997). Design, construction and performance of soil nailed excavation in Hong Kong. In Proceedings of the XIVth International Conference of Soil Mechanics and Foundation Engineering, Hamburg, Germany, 6–12 September, pp. 1339–1342.

Smith, I.M. & Su, N. (1997). Three-dimensional FE analysis of a nailed wall curved in plan. *International Journal for Numerical and Analytical Methods in Geomechanics*, vol. 21, pp. 583–597.

Stocker, M.F. & Reidinger, G. (1990). The bearing behaviour of nailed retaining wall. In Proceedings of Design and Performance of Earth Structure, Geotechnical Special Publication No. 25 (edited by Lambe and Hansen), ASCE, pp. 613–628.

Tan, S.A., Luo, S.Q. & Yong, K.Y. (2000). Simplified models for soil-nail lateral interaction. *Proceedings of the Institution of Civil Engineers: Ground Improvement*, vol. 4, no. 4, pp. 141–152.

Tayama, S. & Kawai, Y. (1999). The current status and outlook for soil nailing in Japan. In Special volume for the Proceedings of the 11th Asian Regional Conference on Soil Mechanics and Geotechnical Engineering, Korea, pp. 51–61.

Tei, K., Taylor, R.N. & Milligan, G.W.E. (1998). Centrifuge model tests of nailed soil slopes. *Soils and Foundation*, vol. 38, no. 2, pp. 165–177.

Thompson, S.R. & Miller, I. R. (1990). Design, construction and performance of a soil nailed wall in Seattle, Washington. In Proceedings of a Conference on Design and Performance of Earth Retaining Structures, Geotechnical Special Publication No. 25, New York, 18–21 June, pp. 629–643.

Zhang, J., Zhang, L.M. & Tang, W.H. (2011). New methods for system reliability analysis of soil slopes. *Canadian Geotechnical Journal*, vol. 48, pp. 1138–1148.

Zhou, Z. (2008). Centrifuge and Three-Dimensional Numerical Modelling of Steep CDG Slopes Reinforced with Different Sizes of Nail Heads. PhD thesis, The Hong Kong University of Science and Technology, Hong Kong.

Zolqadr, E, Yasrobi, S.S. & Olyaei, M.N. (2016). Analysis of soil nail walls performance: Case study. *Geomechanics and Geoengineering*, vol. 11, no. 1, pp. 1–12.

Chapter 3

Site investigation

3.1 INTRODUCTION

Site investigation is the process of collecting information and evaluating the site conditions for the purposes of designing and executing a civil engineering or building project. It is generally carried out at the early stage of an engineering project. Site investigation is an essential and critical part of a project which facilitates the designer to appreciate the geological, hydrogeological, and environmental conditions of a site such that the geotechnical risk of the proposed works throughout the design, construction, and maintenance stages could be identified, and minimized or eliminated. Nevertheless, it is not uncommon for the value of site investigation not to be fully understood until construction problems or failures due to unforeseen ground conditions occur. The lack of a thorough site investigation plan may lead to various undesirable outcomes, including uneconomical designs, need for re-designs, construction difficulties, additional construction costs and delays, and inadequate or unsafe performance of the works. The risk associated with inherent uncertain ground conditions has led to cost and time overruns on many construction projects. According to European statistics, about 80% to 85% of all building failures and damages are related to unforeseen and/or unfavorable ground conditions. In contrast, a well planned and executed site investigation has a direct positive effect on the feasibility and cost-effectiveness of a project by minimizing the chance of over-conservative design, re-design, or delay due to unforeseen ground conditions.

As for other civil engineering and building projects, it is important to define clearly the purpose of a site investigation for a soil nailing project before its commencement. In general, the aim of a site investigation is to obtain adequate information about the ground mass to be reinforced, by the most appropriate and economical means, so as to allow safe, economical, and environmental design and construction. An adequate site investigation could help to establish the geological and hydrogeological model of the site, establish the general suitability of applying the soil nailing technique, determine pertinent design parameters, and identify potential problems

Table 3.1 Objectives of a site investigation

Objectives	Description
Feasibility	To assess the technical feasibility of applying the soil nailing technique to the site.
Design Consideration	To enable a safe, economical, and environmental design to be prepared, including the design of temporary works.
Construction Consideration	To plan the best method of construction; to foresee and formulate possible solutions against difficulties and delays that may arise during construction due to adverse ground conditions and other factors in appropriate cases; to explore sources of suitable materials for use in construction; to select sites for the disposal of waste or surplus materials, etc.
Effects of Changes	To determine the changes that may arise due to the ground and environmental conditions, either naturally or as a result of the works, and the effect of such changes on the works, on adjacent works, and on the environment in general.
Site Selection	Where alternatives exist, to advise on the relative suitability of different sites, or different parts of a given site.

that may be encountered during construction or maintenance stage. The key objectives of a site investigation are summarized in Table 3.1.

References should be made to Clouterre (1991), ICE (2012), Eurocode 7—Part 1 (EN 1997-1:2004), Lazarte et al. (2015), and GEO (2017a & b) for a further discussion of the design and execution of site investigation. Only pertinent aspects of site investigation, especially those related to the design and construction of soil nailing works, are summarized in this chapter.

3.2 SCOPE AND SEQUENCE OF SITE INVESTIGATION

3.2.1 General

There are two major components of a site investigation, namely general appraisal and subsurface investigation. General appraisal mainly involves examination of the site setting and available records through desk study and field inspection, while subsurface investigation is usually conducted through subsurface exploration. The planning of subsurface exploration should be based largely on the findings of the general appraisal. A site investigation will normally proceed in stages, viz. (i) desk study, (ii) site reconnaissance, (iii) subsurface exploration, including ground investigation and laboratory testing, and (iv) follow-up investigation for design review during construction, if needed.

The cost of a site investigation is generally low in comparison with the overall cost of a project and may be further reduced by diligent forward planning. Discussion at an early stage with specialist contractors can help to formulate an appropriate and economical site investigation plan. The technical requirements of the investigation should be the overriding factor in the selection of investigation methods, rather than their costs.

3.2.2 Desk study

As the first stage in a site investigation, a desk study is carried out to examine all available regional and site-specific geotechnical, geological, and hydrogeological information pertaining to the project site, including but not limited to the following:

- Topographic and survey maps.
- Geological maps and the associated information including lithology of rock outcrops, landform, erosion pattern, past landslides, etc.
- Aerial photographs.
- Geotechnical data and investigation reports of nearby sites.
- Meteorological and hydrological information.
- Information on the performance of nearby structures.
- Records of existing services and utilities.
- Environmental constraints.
- Lease conditions.
- Information on seismicity.

Upon collection of the information, a review of their quality, reliability, validity, and applicability to the proposed project should be carried out. The assembly of the desk study information should be completed as far as possible, at least with respect to ground conditions, before the ground investigation works begin. A preliminary ground investigation may sometimes be desirable to help determine the extent and nature of the main ground investigation.

3.2.3 Site reconnaissance

Site reconnaissance consists of visual inspection of the site and collection of pertinent data. At an early stage, a thorough visual examination should be made of the site to identify the key topographical and geological features, as well as to appreciate the access and working conditions of the site. In addition, an examination of the regional topography and drainage pattern may also provide valuable information about the probable subsurface structure of the site. The extent to which the ground adjacent to the site should be examined is, in essence, a matter of judgment. In general, the following information should be obtained during the site reconnaissance:

- General site conditions.
- Geology and geomorphology, including rock outcrops and landforms.

- Site accessibility for work forces and equipment, in particular for site exploration and construction.
- Equipment storage and security.
- Adjacent land use.
- Traffic condition.
- Surface drainage patterns.
- Nature and condition of nearby structures.
- Nature and condition of underground and above-ground services and utilities.
- Evidence of surface settlement and erosion.
- Evidence of unfavorable soil or groundwater conditions.
- Evidence of corrosion in nearby structures.
- Easement requirements.

3.2.4 Subsurface exploration

The purpose of subsurface exploration is to ascertain and characterize the ground conditions and groundwater regime in the area that would affect or be affected by the proposed soil nailing works. This includes determination of the subsurface stratigraphy and the engineering properties of the geological materials underlying the site through field and laboratory testing, assessment of the ground and groundwater corrosion potential, and evaluation and monitoring of the groundwater condition. In general, the planning of a subsurface exploration program involves determining the number, locations and depths of boreholes and trial pits, formulating the strategy and methods of sampling, assessing the types of field and laboratory testing to be carried out, and the requirements for appropriate equipment and instrumentation. A subsurface exploration normally consists of (i) *in situ* testing of soil/rock properties, (ii) retrieval of representative samples of soil/rock for direct mapping and laboratory testing, (iii) characterization of the stratigraphy, and (iv) identification and monitoring of the pertinent environmental factors such as groundwater conditions.

The ground profile and stratigraphy should be established in a representative geological model that enables adequate characterization of the lateral and vertical extent of all geological materials and their nature. The principal engineering properties of interest include strength, deformation, and hydraulic characteristics of the ground materials. The basic characteristics and properties to be determined include, but not limited to, physical properties such as index parameters, shear strength, compressibility, and corrosion potential, as well as phreatic and piezometric levels. Special attention should be given to identify any adverse geological features and adverse groundwater conditions that may affect the application of soil nailing.

Apart from the conventional techniques for subsurface exploration, one may consider adopting geophysical tests for stratigraphic profiling and delineation of underground features. This may include measurement of seismic waves (e.g. seismic refraction survey, cross-hole, downhole, or spectral analysis of surface wave test, etc.) and electromagnetic techniques (e.g. resistivity, EM, magnetometer, radar, etc.).

Subsurface exploration may either be conducted in a single phase or in different stages, depending on the complexity of the project and the availability of existing geotechnical data. If little or no geotechnical information is available for a site, conducting the investigation in two phases may be a sensible approach. The first phase would consist of a preliminary subsurface investigation to help identify the soil strata and their basic characteristics, and to determine if the ground is favorable, difficult, or unfavorable for soil nails. The second phase would consist of a detailed subsurface investigation focused on collecting the necessary data required for design and construction of the soil nailed structure. There is no simple way to define the extent and scope of subsurface exploration. It depends on the geological conditions of the site, the risk level involved, size and complexity of the soil nailing project, time and budget available, and engineering judgment. Nevertheless, there are plenty of general guidelines available in the literature for reference (e.g. Lazarte et al., 2015; GEO, 2017a & b).

3.2.5 Follow-up investigation for design review during construction

Notwithstanding good planning and proper execution of a site investigation, it can be difficult to forecast the ground conditions from a limited available information because of the inherent variability and heterogeneity of the geological materials. Occasionally, new information is revealed during the construction stage which may necessitate follow-up investigation and design review. In some cases, this may trigger the need to amend the design or construction procedures.

3.3 SUBSURFACE EXPLORATION TECHNIQUES

3.3.1 General

The appropriate techniques to be displayed for the subsurface exploration program should be planned carefully by experienced geotechnical professionals. There are various subsurface exploration techniques available in the market. The selection of a suitable technique depends on the type and accuracy of the information to be obtained, ground conditions, site constraints, accessibility, time, and budget. The most commonly used techniques include trial pits, trench excavation, boreholes, *in situ* testing, geophysical testing, etc.

3.3.2 Trial pits and trench excavation

Shallow trial pits permit direct visual inspection of the soil and rock formation without the need for any special equipment. They are commonly dug manually to a depth of about 3 m. It is essential to ensure that the sides of excavation are well-guarded against collapse and meet the required safety standards. The technique allows detailed examination of the *in situ* ground conditions and their variability both vertically and laterally. They also facilitate retrieval of good quality samples for *in situ* or laboratory testing. Apart from assessing the ground conditions, the technique may also be used to ascertain the location of buried utilities and services, and to investigate the geometry of underground structures. Trial pits could also be extended to trenches or slope surface stripping to suit the site conditions and the information required. This is particularly useful when soil nailing is used as a remedial measure to stabilize failed slopes where shear surfaces or potential shearing zones could be located.

Another benefit of adopting the technique of trial pitting or trenching is that it provides a quick and economical method that helps assess the ability of an excavated face to stand while temporarily unsupported during the staged excavation or formation of boreholes for soil nails. This is important for assessing the buildability of a soil nailed structure.

3.3.3 Boreholes

There are various methods for advancing boreholes into the ground. The technique allows retrieved samples for visual inspection, *in situ* and laboratory testing, as well as installation of instrumentation such as standpipes and piezometers. There is, however, no "cookbook-type" exploration plan to suit all sites. The plan, which usually includes the number, location, and depth of boreholes, type of test to be carried out, and the type of instrument to be installed, should be developed based on experience and judgment with due regard to the information available from the desk study and site reconnaissance, complexity and variability of the ground conditions, time and cost constraints, etc.

In general, boring and drilling are the commonly used methods for borehole advancement. Boring is suitable in relatively soft and uncemented ground materials such as soils and soft rocks, which are normally close to the ground surface. The most common techniques in this category include hand augering, mechanical augering, and light percussion boring. Drilling, on the other hand, is suitable for more competent or cemented materials, which are normally situated at greater depths. The techniques under this category include rotary open hole drilling and rotary core drilling. These make use of a rotary action combined with downward force to grind away the materials and form a hole. The techniques can be applied to both soils and rocks. Core samples can be recovered for inspection and testing where needed. The drilling rigs are mostly of the hydraulic feed type where the flushing medium may be water, air, air-foam, or mud. By far, this is the most commonly used investigation method in many places of the world.

3.3.4 Probing

There is an increasing trend of using probing as a supplement to boreholes for subsurface exploration. A wide range of static and dynamic probing techniques are available to suit different ground conditions. For example, a hand-held dynamic probe, known as the GCO probe, is commonly used in Hong Kong, whereby a 10 kg hammer is dropped by a distance of 300 mm. The blow counts for each 100 mm penetration could then be related to the engineering properties of the ground materials based on established correlations. The technique is rapid and economical, which only requires simple lightweight equipment and is particularly valuable for ground profiling. A drawback of the technique is that it cannot be used to obtain samples or to install instrumentation. In general, the technique is suitable for application in most of the ground conditions except in very stiff or very dense soils.

3.4 TESTING AND MONITORING

3.4.1 *In situ* field testing

It is common that field tests are carried out as part of the ground investigation. Field tests enable the strength, deformation, and permeability characteristics of both soils and rocks to be assessed. This facilitates a more realistic appraisal of the ground conditions on a larger scale when compared with laboratory testing. The index properties measured by the various field tests can be related to the engineering properties of the soils based on established correlations. In general, *in situ* field testing includes penetration test (e.g. standard penetration test (SPT), cone penetration test (CPT)), shear strength test (e.g. vane shear test), pressuremeter test, impression packer survey, plate load test, permeability test, and geophysical test (e.g. seismic, resistivity, electromagnetic, magnetic, gravity).

In situ field testing is particularly useful when it is difficult or expensive to obtain good quality undisturbed samples for laboratory testing. For example, it is inevitable that conventional sampling of granular soils would cause some irreversible disturbance and change the *in situ* density and strength of the soils. The use of field testing such as SPT and CPT to correlate the density and shear strength of the soils for design would be more representative. There are plenty of guidance documents available on *in situ* field testing (e.g. ICE, 2012; GEO, 2017a). Readers are recommended to refer to these documents for further details.

3.4.2 Laboratory testing

Laboratory testing will characterize the ground materials and help to confirm the feasibility of applying the soil nailing technique to a site. It will also provide a basis for determining the necessary engineering parameters for the soil nail design. However, it is essential to ensure that laboratory tests are carried out on these samples of soils and rocks that are sufficiently representative of the ground materials at the site. Tests commonly carried out include classification tests, measurement of density, grading, permeability and shear strength, as well as the corrosion potential of the ground materials and groundwater. Both undrained and drained shear strength tests should be conducted to determine the short-term and long-term behavior of the soils as appropriate. Sometimes, direct shear test may be conducted to determine shear strength at large strains. This can provide useful information on the shear strength along planes of weakness such as shear planes, joints, and clay seams. If ground settlement is of concern, in particular when the proposed soil nailed structure is built in soft or loose soils, consolidation test using oedometer or hydraulic apparatus should be carried out. Swelling test may also be conducted to quantify swelling pressure that might arise due to unloading causing degradation of the slopes in plastic clay.

3.4.3 Pullout test

Pullout test may be carried out during the site investigation stage to determine the ultimate pullout resistance of soil nails for the subsequent design. It also helps to verify the suitability of the proposed construction procedure and workmanship. This is a cost-effective approach to obtain design information on site-specific pullout resistance for large-scale projects or for sites where there is relatively little experience with the behavior of the ground. In general, the number of investigation pullout tests will depend on the relevant soil strata into which production soil nails are to be bonded.

The investigation pullout test is normally carried out on non-production (sacrificial) nails. These test nails should be installed using the same procedures as for the production soil nails except that only the part of the soil nail within the specific layer, for which the bond information is required, is to be grouted, and other parts are to be de-bonded from the ground. This is to ensure the test nails are the representative of the production nails and that the bond strength is obtained from the relevant zone. More details about investigation pullout test are given in Chapter 6.

3.4.4 Groundwater monitoring

The design groundwater level is one of the major uncertainties in the design of a soil nailed structure, in particular if the groundwater level fluctuates within the reinforced soil mass. It is also a major factor that affects the suitability of applying soil nailing at a site as well as the long-term stability of the soil nailed structure. It is therefore essential that the location and fluctuation of the groundwater levels as well as any perched water conditions in the surface mantle should be ascertained during the site investigation stage. Normally, standpipes and piezometers are installed in boreholes as part of the site investigation for groundwater monitoring. Groundwater seepage and surrounding water levels should be observed and recorded during the advancement of the borehole. It is preferable that the groundwater conditions could be monitored for a suitably long period of time, e.g. more than 12 months, such that information on the seasonal fluctuation could be collected for subsequent analysis. In practice, however, this may not be possible due to tight construction program. Thus, due consideration should be given to the published local rainfall data and the long-term trend, so as to derive a sufficiently reliable design groundwater level for the service life of a soil nailed structure. For example, more than 120 rain gauges are available in Hong Kong with a total land area of about 1,100 km² which have provided a long period dataset (ranging from 20 to more than 100 years of record) to establish a reliable rainfall intensity-duration correlation for different rainfall return periods. With the rainfall data and site-specific groundwater measurements, a reasonable estimate of the design groundwater level could be obtained. Nonetheless, climate change effects can lead to the more frequent occurrence of extreme rainstorm events. For example, the record for hourly rainfall in Hong Kong has been broken several times in the past few decades, whereas it used to take several decades to break the record in the past (see Figure 3.1). One should therefore make due allowance for the effect of extreme rainfall in estimating the design groundwater level where considered appropriate.

3.4.5 Chemical test

Durability is one of the key design requirements of a soil nailed structure. The acidity of ground materials and groundwater poses different degrees of corrosion potential to steel soil nails. It is important to appreciate the aggressivity of a site in order to provide appropriate corrosion protection measures to the soil nails. The aggressivity of soils can vary over a wide range because of the heterogeneity and variability of soil compositions and properties, and other environmental factors.

In general, the corrosion rate of steel soil nails is affected by the physical and chemical characteristics of the soil within which the soil nails are embedded. Pertinent physical characteristics include those that control the

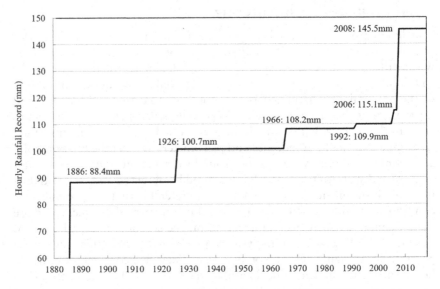

Figure 3.1 Hourly rainfall records in Hong Kong from year 1885 to 2019.

permeability of the soil for the passage of air and water. Fine-grained soils, i.e. silts and clays, are potentially more corrosive than coarse-grained soils, i.e. sands and gravels, in which there is a greater circulation of air and less water retention capacity. Pertinent chemical characteristics include those that determine the ability of the soil to act as an electrolyte for the development of local corrosion cells. These include alkalinity, acidity, concentrations of oxygen and dissolved salts, organic matter, and bacteria content. A detailed assessment of soil aggressivity should be made by means of physical and chemical testing in the laboratory, review of site records, and field observations. More details about soil aggressivity assessment are given in Chapter 5.

3.5 BUILDABILITY OF SOIL NAILS

One of the key objectives of a site investigation for a soil nailing project is to assess the buildability of soil nails in order to ensure that the design is practical and buildable. The buildability of soil nails is, to a large extent, governed by the ground and groundwater conditions. Certain ground conditions are more likely to present problems for soil nail construction (also see Table 1.3 in Chapter 1). For example, the chance of encountering potential problems of excessive grout leakage during soil nail installation is higher if the geological conditions comprise generally permeable coarse materials with a relatively low silt/clay content in the matrix and moderate to high intergranular porosity, or where geological structures are present that promote enhanced fluid throughflow. The following geological conditions are susceptible to excessive grout leakage during soil nail installation:

- Fill materials containing a significant proportion of coarse materials, i.e. boulders, cobbles, gravels, and sands.
- Colluvium and fluvial deposits with a high proportion of coarse materials.
- Soil erosion pipes that may be partly infilled by porous and permeable materials.
- Material boundaries within colluvium of different compositions, and between colluvium and *in situ* material, and within corestone-bearing saprolite (i.e. completely or highly weathered rocks), especially at the margins of corestones, open joints, faults, and shear zones, and other discontinuities (e.g. zones of hydrothermal alternation) that are weathered and eroded, and hence are open and act as preferential flow paths.
- Landslide scars, tension cracks, and other distress features due to slope deformation, as these may include voids within transported and in situ materials.
- Buried drainage lines that intersect slopes and may contain loose colluvium, erosion pipes may develop, and preferential groundwater throughflow as indicated by seepage locations/horizons may also occur.

Any leaked grout during soil nail installation could permeate into voids present in the ground and potentially dam up the groundwater flow in the vicinity of the soil nailed structure. Sufficient information about the ground and groundwater conditions should be collected for assessing the buildability of a soil nailed structure. This information also provides the basis for the formulation of ground models for the design of a soil nailed structure. If there is a concern about potential damming up of groundwater due to soil nail construction, piezometers should be installed at appropriate locations after completion of the works and monitored for a sufficient period of post-construction time in order to ascertain such potential.

Drilling for long soil nails, typically over 20 m, stands a higher chance of intersecting local or perched groundwater tables and adverse geological features, such as local weak geological zones or seams, and dykes of high hydraulic permeability contrast. This may lead to construction problems such as collapses of soil or rock fragments along the borehole and large amounts of grout leakage, which in turn can adversely affect the quality of soil nails. For cases where long or closely spaced soil nails are proposed, or where the ground or groundwater conditions are likely to be adverse to soil nail construction, designers should consider undertaking an assessment of buildability and the effects of soil nail installation on the existing ground and groundwater conditions. This may include the construction of trial nails prior to carrying out the soil nailing works. By judicious positioning of the trial nails, the site trial can provide useful information on potential construction problems, such as areas of potential excessive grout leakage, presence of loose materials prone to hole collapse, high groundwater levels, excessive ground movement, and ground loss that may impose on adjacent facilities or services, etc. This information is important for assessing the buildability of soil nailing works. It also facilitates better design of working soil nails, and planning of appropriate precautionary measures to overcome the site problems. Details of the trial including its locations, potential problems encountered, and the necessary contingency or precautionary measures should be included in the designer's requirements under the contract.

It is important to have a thorough understanding and appreciation of the geological and hydro-geological conditions of the site in order to assess the buildability of soil nails. The site investigation should be sufficiently detailed to affirm the buildability of the proposed soil nailing works and to obtain information for detailed design. The investigation should not be confined to the ground within which soil nails are to be embedded. The ground mass in the vicinity of the soil nailed structure, which may affect the overall stability and deformation of the proposed soil nailed structure, should also be investigated.

3.6 CONCLUDING REMARKS

Good planning and proper execution of a site investigation is essential to the successful completion of a soil nailed earth structure. The site investigation should be carried out at the early stage of an engineering project. With the essential information on the geological, hydrogeological, and environmental conditions of a site, the geotechnical risk posed by the proposed soil nailing works throughout the life cycle, including the design, construction, and maintenance stages, could be identified, and minimized or eliminated. In this chapter, the purpose and scope of a site investigation for soil nailing are discussed. A typical site investigation includes desk study, site reconnaissance, and subsurface exploration, and in some cases, follow-up investigation for design review during the construction stage. Various ground investigation techniques and testing methods are introduced. Readers should bear in mind that a well-planned and executed site investigation is essential for formulating representative geological and hydrogeological models, and determining suitable design parameters for the site. It also facilitates the designer to assess the suitability and buildability of the soil nailing technique for the project.

REFERENCES

Clouterre (1991). French National Research Project Clouterre: Recommendations Clouterre (English Translation 1993). Federal Highway Administration, US Department of Transportation, Washington, D.C., USA, Report No. FHWA-SA-93-026, 321 p.

EN 1997-1 (2004). Eurocode 7: Geotechnical design: Part 1: General rules, 168 p.

GEO (Geotechnical Engineering Office) (2017a). *Guide to Site Investigation (Geoguide 2)*. Geotechnical Engineering Office, Civil Engineering and Development Department, HKSAR Government, Hong Kong, 349 p.

GEO (2017b). *Model Specification for Soil Testing (Geospec 3)*. Geotechnical Engineering Office, Civil Engineering and Development Department, HKSAR Government, Hong Kong, 324 p.

ICE (Institution of Civil Engineers) (2012). *ICE Manual of Geotechnical Engineering – Volume 1: Geotechnical Engineering Principles, Problematic Soils and Site Investigation* (edited by J. Burland, T. Chapman, H. Skinner & M. Brown). Institution of Civil Engineers, London, UK, 727 p.

Lazarte, C.A., Robinson, H., Gomez, J.E., Baxter, A., Cadden, A. & Berg, R. (2015). Geotechnical Engineering Circular No. 7: Soil Nail Walls: Reference Manual. Federal Highway Administration, US Department of Transportation, Washington, DC, USA, Report No. FHWA-NHI-14-007, 385 p.

Chapter 4

Design and analysis

4.1 INTRODUCTION

This chapter outlines the design process of a soil nailed structure. The objective of the design of a soil nailed structure is to arrive at the required layout and dimensions of soil nails such that the reinforced structure can perform its intended functions in ways that are economical, safe, and environmentally friendly throughout the design life or service life. The design of a soil nailed structure usually starts with preliminary design at the early stage, followed by detailed design. During the preliminary design stage, one should establish the design requirements, identify the potential failure modes, loading conditions, and site constraints, assess buildability, and determine the preliminary layout and dimensions of soil nails. In the subsequent detailed design stage, the designer should collate site-specific geotechnical, geological, and hydrogeological information through site investigation, and make use of available analytical tools to design the soil nailed structure to ensure compliance with safety, serviceability, and durability requirements, as well as other relevant project requirements. The design process should, however, continue throughout the construction stage whereby the original design may be suitably revised to cater for new information collected from field observations. The key elements of the design process are illustrated in Figure 4.1. Worked examples illustrating the key steps of the design of a soil nailed structure, both using the approaches of Allowable Stress Design (or Working Stress Design) and Limit State Design (or Load and Resistance Factor Design) are given in Appendices A and B, respectively. Readers are recommended to read through the example in conjunction with this chapter for a better understanding of the design process.

4.2 PRELIMINARY DESIGN

4.2.1 General

Before undertaking a preliminary design, the designer should be satisfied that soil nailing is the most suitable design scheme amongst other options

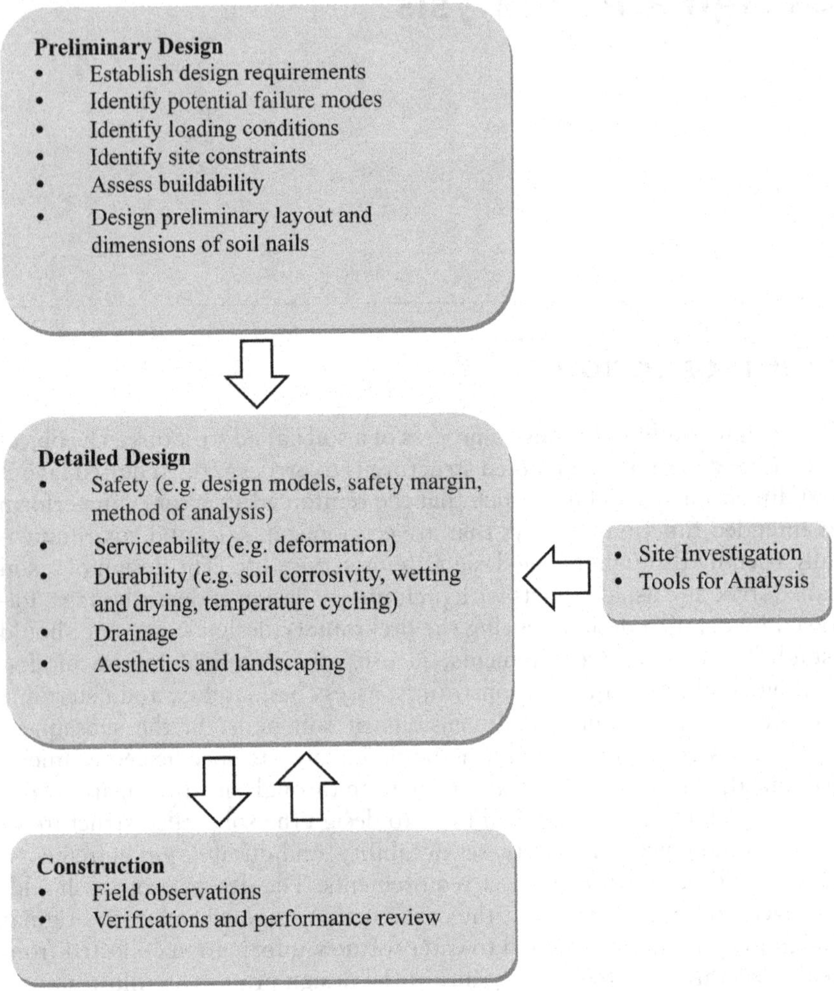

Preliminary Design
- Establish design requirements
- Identify potential failure modes
- Identify loading conditions
- Identify site constraints
- Assess buildability
- Design preliminary layout and dimensions of soil nails

Detailed Design
- Safety (e.g. design models, safety margin, method of analysis)
- Serviceability (e.g. deformation)
- Durability (e.g. soil corrosivity, wetting and drying, temperature cycling)
- Drainage
- Aesthetics and landscaping

- Site Investigation
- Tools for Analysis

Construction
- Field observations
- Verifications and performance review

Figure 4.1 Design process of a soil nailed structure.

from the perspective of technical, environmental, economic, and program considerations. A preliminary design is basically a bridge between design conception and detailed design, where the level of conceptualization achieved during the ideation stage is usually not quite sufficient for full evaluation. The preliminary design is an iterative process in which the design intents are specified and various engineering solutions in respect of using the soil nailing technique may be conceived. This stage of design involves a thorough appreciation of the engineering challenges and constraints, and evaluation of different options subject to the design requirements. The feasibility of adopting a soil nailed structure should also be assessed. In the following sections, the key components involved in a preliminary design, namely design requirements, failure modes, loading conditions, site constraints, buildability and preliminary layout, and dimensions of a soil nailed structure, are further elaborated.

4.2.2 Design requirements

Like other civil engineering structures, a soil nailed structure is required to fulfil a set of design requirements including design life, stability, serviceability, and durability throughout the life cycle (i.e. including construction stage and service life) of the structure. Other issues such as construction safety, cost, and environmental impact are also important design considerations. Table 4.1 summarizes the common key requirements for designing a soil nailed structure.

4.2.3 Modes of failure

Because a soil nailed structure consists of various components ranging from soil, rock, to structural elements such as reinforcement, nail heads, and facing, each component may be associated with one or more potential failure modes. It is of paramount importance to identify all the potential modes of failure under specific geological and hydrogeological conditions. The failure modes of a soil nailed structure can be broadly classified into two categories, namely (i) external failure modes and (ii) internal failure modes. External failure refers to the global instability of the structure, whereas internal failure refers to instability of one or more components. Figures 4.2 and 4.3 illustrate some common modes of failure that should be considered in the design of a soil nailed structure.

4.2.3.1 External failure mechanisms

External failure is associated with the development of failure surfaces essentially outside the soil nailed reinforced ground mass (i.e. global ground instability). The evaluation of external failure is an important aspect of the design process as the magnitude of the failure could be large and hence the consequence can be significant or serious. For external failure modes, the soil nailed ground mass is generally considered as a reinforced block whereby the failure can be in the form of sliding, rotation, bearing, or other forms of loss of overall stability as illustrated in Figure 4.2.

If some of the soil nails are intersected by the potential failure surface, the resistance provided by the intersected soil nails should be considered in the evaluation of overall stability of the soil nailed structure. However, the potential internal failure mechanism should also be considered for these intersected soil nails (see below). For marginally stable steep slopes, retaining walls, or embankments, which have been assessed and found to require

Table 4.1 Common key design requirements of a soil nailed structure

Design Requirements	Details
Design Life	The design life of a soil nailed structure is the period of time that the structure can perform its intended purposes properly and safely without unexpected costs or disruption due to maintenance and repairs. The design life depends on the permanence of the soil nailed structure (i.e. temporary or permanent works). It should be clearly specified at the early stage of a design as the design life affects the safety level and hence the cost required. For most applications, a permanent soil nailed structure has a service life of about 50 to 120 years. Temporary applications, however, have a much shorter service life, typically less than 2 to 3 years. Occasionally, the safety level required for temporary applications may be equivalent to that for permanent applications. An example is an excavation for basement construction in a densely developed urban area. If the service life of a temporary soil nailed structure has to be extended, an assessment in respect of the relevant design requirements including stability, durability, etc., should be carried out to determine whether additional engineering measures are necessary.
Stability	The stability of a soil nailed structure during construction and throughout its service life should be assessed and ensured. Its performance should not exceed a state at which failure mechanisms can form in the ground or within the soil nailed structure, or when deformation of the soil nailed structure can lead to severe damage to its structural elements or nearby structures, facilities, or services. The design of a soil nailed structure should ensure that there would be adequate safety margins against all the perceived potential modes of soil and reinforcement failure. The required safety margins depend on the risks posed to life and property as well as the failure mechanisms. For example, Eurocode EN 1997-1 requires one class of safety margin only regardless of the risk level, whereas in Hong Kong and Sweden there are three classes of safety margin for slope improvement design, depending on the consequence of failure.
Serviceability	The mobilization of tensile force in soil nails necessitates ground deformation. The performance of a soil nailed structure should not exceed a state at which the deformation of the structure affects its appearance or the efficient use of nearby structures, facilities, or service. Potential serviceability problem associated with soil nailed structures commonly involves excessive ground deformations.
Durability	Durability is a major consideration in a soil nailed structure. The environmental conditions should be investigated at the design stage to assess their significance to the durability of soil nails. Appropriate measures should be applied to the soil nails such that an adequate safety margin of the soil nailed structure can be maintained throughout its service life. The durability of a soil nailed structure is governed primarily by the resistance of reinforcement to corrosion due to chemical attack, wetting and drying, prolonged stress application, and temperature cycling.

(Continued)

Table 4.1 (Continued) Common key design requirements of a soil nailed structure

Design Requirements	Details
Drainage	Provision of effective drainage systems, both surface and subsurface drainage, is essential to avoid the build-up of excessive water pressures within and behind a soil nailed structure, and hence enhance its satisfactory performance during the construction and service life. High redundancy and minimal effort of maintenance should be considered in the design of the drainage systems.
Environment/ Aesthetics	The construction of a soil nailed structure may disturb the ground ecosystem, induce nuisance and pollution during construction, and cause visual impact to the existing environment. Adverse impact to the environment should be minimized. For example, mature trees and natural terrain should be preserved and protected whenever possible in order to sustain the ecosystem. Appropriate pollution control measures, such as provision of water sprays and dust traps at the mouths of boreholes when drilling into rocks and boulders, screening the temporary working platform, and installing noise barriers in areas with sensitive receivers, should be undertaken. Hard slope cover should be adopted as a last resort; where considered necessary, it can be dyed, sculpted, or molded to replicate the appearance of natural rock formations, stone, or brickwork.
Safety during Construction and Maintenance	The most cost-effective and practical approach to avoid introducing hazards to a soil nailed structure throughout its service life, including the construction stage is by eliminating them during the early design stage. The potential hazards should be identified such that preventive and mitigation measures can be developed and implemented during the construction period and, if necessary, throughout the service life of the structure.
Cost-effectiveness	The construction as well as maintenance costs of a soil nailed structure depend on a number of factors including material cost, construction method, temporary works requirements, buildability, corrosion protection requirements, soil nail layout, type of facing, etc. This should be duly considered during the preliminary design stage. As good practice, one should estimate and consider the whole life cycle costing in the early stage of design.

only a few widely spaced soil nails, the concept of treating the reinforced block as a monolith may not be appropriate.

4.2.3.2 Internal failure mechanisms

Internal failure refers to the development of failure mechanisms within the soil nailed ground mass. This is essentially related to the failure in load transfer between the ground and the soil nailing components. Internal failures can occur in the active zone, passive zone, or in both the active and passive zones of a soil nailed structure as illustrated in Figure 4.3.

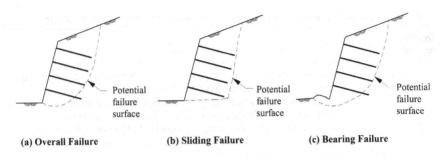

Figure 4.2 Potential external failure modes of a soil nailed structure (GEO, 2008).

Typical internal failure modes in the active zone include:

- Failure of the ground mass, i.e. the ground disintegrates and "flows" around the soil nails and soil nail heads.
- Bearing failure underneath soil nail heads.
- Structural failure of soil nail under combined actions of tension, shear, and bending.
- Pullout failure at ground-grout interface or grout-reinforcement interface for grouted nails, or ground-reinforcement interface if driven or ballistic nails are used.
- Structural failure of the soil nail head or facing, i.e. bending or punching shear failure, or failure at head-reinforcement or facing-reinforcement connection.
- Surface failure between soil nail heads, i.e. washout, erosion, or local sliding failure.

Typical internal failure modes in the passive zone include pullout failure at ground-grout interface or grout-reinforcement interface for grouted nails, and ground-reinforcement interface for driven or ballistic nails. Pullout failure is indeed a primary failure mode in a soil nailed structure. The ability of a soil nail to provide adequate pullout resistance depends on a number of factors such as soil type, groundwater conditions, drilling and installation technique, surface roughness of drillhole, grout annulus and reinforcement, diameter of drillhole, and grouting pressure. Various methods have been put forward to assess the ultimate pullout resistance of soil nails (see Chapter 6 for more details on pullout resistance).

4.2.4 Loading conditions

A soil nailed structure may experience simultaneous imposition of different types and combinations of loading during its service life. The loads that should be considered in a soil nailed structure design are generally classified

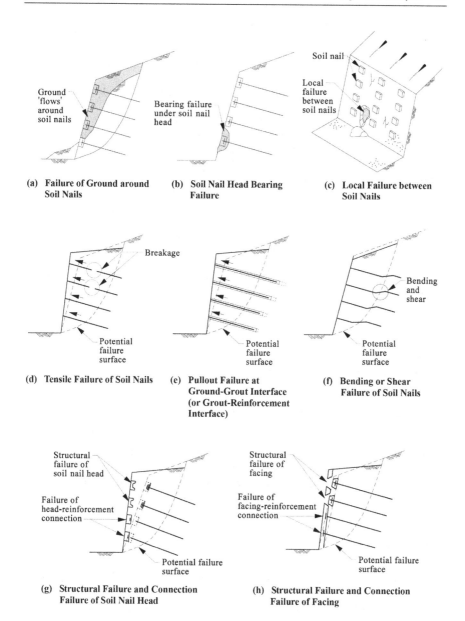

Figure 4.3 Potential internal failure modes of a soil nailed structure (GEO, 2008).

as dead, live, and extreme loads. Typical dead loads include the weight of the soil nailed structure and its attachments, whereas transient loads due to rise in groundwater level and vehicular or pedestrian loads are examples of live loads. Under special circumstances, such as where soil nailing is to be used in cold regions, the soil nailed structure may be exposed to seasonal

Table 4.2 Common loading conditions of a soil nailed structure

Types of Loads	Details
Dead Loads	• weight of structural components such as reinforcement and concrete, • weight of non-structural component such as soil, rock, and water, • earth pressures, and • weight of nearby above-ground structures.
Live Loads	• transient rise in groundwater level (e.g. due to rainfall), • transient traffic load, and • transient load due to freezing temperatures.
Extreme Loads	• seismic loads, • loads due to extreme weather such as extraordinary rise in groundwater level.

ambient freezing temperatures which can cause frost action in saturated soils and the corresponding transient pressures on the facing should be considered. In areas that are prone to earthquake, the design seismic load could be an extreme load to be considered. Arising from the potential impact of climate change, it may also be necessary to consider loading conditions, such as an extraordinary rise in groundwater level induced by extreme weather. Table 4.2 summarizes some common types of loading that should be considered in a soil nailed structure design.

4.2.5 Site constraints and buildability

An adequate appreciation of site constraints to the construction of a soil nailed structure is crucial in assessing the buildability of soil nailing at a given site. Without proper recognition of the site constrains during the early stage of design, an inappropriate soil nailing scheme may be conceived that can subsequently prove impractical or excessively costly in order to cater for the constraints.

4.2.5.1 Access

The "drill-and-grout" soil nailing technique is simple and versatile, rendering it adaptable to physical constraints commonly encountered in typical project sites. In general, the soil nailing technique only requires small-scale construction plant and limited working space. However, there may be situations whereby certain project sites cannot provide adequate working space for soil nail installation. An example is a roadside slope situated in close proximity to a carriageway with heavy traffic density or an infrastructure where there is no room for erection of a temporary working platform for the soil nailing works. Another example is a slope situated in a cliff or a steep coastal area where erection of temporary working platform is extremely difficult. This may pose site constraints to the feasibility of adopting a soil nailing scheme.

4.2.5.2 Ground and groundwater conditions

Successful past project experiences worldwide indicate that the soil nailing technique is suitable for a wide range of soil types and ground conditions. Nevertheless, not all soil types are conducive to soil nailing. Certain ground conditions may lead to problems during the construction stage or in the long term that would make soil nailing too costly or infeasible when compared with other techniques. It is vital for designers to have a good appreciation of the likely ground conditions, including geological and hydrogeological settings of the project site, in the early design stage. Table 4.3 summarizes some common favorable and unfavorable soil types and ground/groundwater conditions for application of the soil nailing technique.

4.2.5.3 Other constraints

Easement may pose a constraint to the application of soil nailing as the soil nails cannot be extended beyond the site boundary to encroach into adjacent land lots. This will usually necessitate the use of more steeply inclined soil nails. However, the resistance that can be mobilized in steeply inclined soil nails will be reduced as compared with shallowly inclined soil nails (see Section 2.2 in Chapter 2). In this case, the designer should evaluate the reduced resistance and ensure that the overall stability of the soil nailed structure would not be compromised. Apart from constraints associated with easement, soil nails will likely sterilize the land above and prevent future development of the site. This should also be considered during the preliminary design stage in option assessment.

Underground obstructions such as live services, existing substructures, and foundations may adversely affect the buildability of soil nailing. This may sometimes preclude the use of the soil nailing technique. The locations of these obstructions should be ascertained as early as possible, so as to enable an assessment of the applicability of soil nailing.

4.2.6 Layout and dimensions of soil nails

In the preliminary design, the layout and dimensions of soil nails, including pattern, inclination, spacing, and length, should be determined. The design depends on a number of factors, such as the strength of the ground (i.e. soil and rock), environmental constraints, geological and hydrogeological settings, and the gradient of the final slope or wall face.

4.2.6.1 Pattern

The arrangement of soil nail pattern is commonly in the form of rectangle, square, or staggered in a triangular pattern (Figure 4.4). In some cases, the soil nails may be in an irregular pattern to suit the site

Table 4.3 Favorable and unfavorable soil types, ground and groundwater conditions for application of the soil nailing technique

Favorable Soil Types/Ground and Groundwater Conditions	Marginally Favorable/Unfavorable Soil Types/Ground and Groundwater Conditions
• Steep or sub-vertical cutting of 1 to 3 m in height can stand in an unsupported manner for at least 1 to 2 days. This will facilitate formation of benches and installation of soil nails in excavations. • Open unsupported drillholes can maintain stability without using casing for at least several hours. This will facilitate insertion of reinforcement and the subsequent grouting works. • Site with a relatively low groundwater table. The most favorable condition is that all soil nails are located above the groundwater table. The excavation faces will be dry which facilitates the drilling and grouting works. • Residual soils and weathered rock without adversely oriented weak discontinuities. This facilitates the grouting works. Weak planes dipping into the excavation may jeopardize the long-term stability of a soil nailed structure. • The degree of weathering of the rock is approximately uniform so that the number of drilling and soil nail installation methods can be minimized. • Glacial outwash and glacial till as they are usually dense and well-graded granular soils with limited fines. • Naturally cemented dense to very dense granular soils with some apparent cohesion. • Firm to stiff fine-grained soils including cohesive soils that are not prone to creep. • Dense engineered fill that is well-graded and placed with an adequate degree of compaction.	• Loose and poorly graded granular soils. The soils may not exhibit adequate stand-up time after excavation. • Ground conditions that necessitate the use of temporary casings during advancement of drillholes. • Highly fractured rock with open joints or voids and open graded coarse granular soils. This will make grouting works difficult because of the potential for excessive grout leakage. • Residual soils and weathered rock with adversely jointing, pre-existing failure surfaces, or compressible soft layers that may cause long-term stability problem to the soil nailed structure. • Soils with a substantial amount of cobbles and boulders. This will make drilling and grouting works difficult. • Sites with a high groundwater table. This may require substantial temporary drainage to facilitate the drilling and grouting works. In the long term, the bond strength at the interface between the grout and the ground may be adversely affected. • Organic or soft fine-grained soils which are susceptible to creep. The bond strength that can be mobilized at the nail-ground interface may be too low. • Highly frost-susceptible soils and expansive soils. • Weathered rock with a highly variable degree of weathering. This may require changes in drilling equipment and installation method, leading to high cost and prolonged construction period. • Loess, which has a low shear strength and may lose its strength significantly upon wetting. This may require long soil nails and drainage measures to prevent water ingress to the reinforced zone. • Soils that are subjected to repeated freeze-thaw process. This may reduce the bond strength at the nail-ground interface in the long term. • Loose granular soils that are susceptible to liquefaction in seismically active regions. • Highly corrosive soils (e.g. cinder and slag), which require expensive corrosion protection measures. • Loose engineered fill that may cause problems during drilling and grouting works. • Non-engineered fill, in particular those with aggressive and loose materials. • Ground with pre-existing failure surfaces.

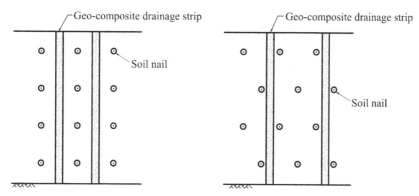

Regular Rectangular / Square Layout   Regular Triangular Layout

Figure 4.4 Common patterns of soil nails.

conditions. The main advantage of adopting a regular pattern is that it can provide a more uniform distribution of resistance to the reinforced ground mass. Furthermore, a regular rectangular or square pattern facilitates the construction of vertical joints in shotcrete facing and geo-composite drainage strips behind the facing of a soil nailed structure. A regular triangular pattern, on the other hand, can enhance the local stability of the face. Some researchers reported that by adopting a triangular nail pattern, the load that a soil nailed structure can sustain would be increased and the corresponding ground deformation would be reduced as compared with that with a regular square or rectangular nail pattern (Singh & Shrivastava, 2017).

4.2.6.2 Inclination

Soil nails are commonly installed at an inclination of 5° to 20° below horizontal and perpendicular to the slope or wall face. Generally speaking, this arrangement facilitates the mobilization of soil nail forces and a more even distribution of resistance to the reinforced zone. For "drill-and-grout" soil nails, the range of this inclination angle also facilitates effective grouting works, in particular when the grouting works are carried out under gravity or low pressures (normally 5 to 10 bar, i.e. 500 to 1,000 kPa). Under special circumstances, such as for avoidance of underground services, foundations or other obstructions, or where significantly stronger ground is located at a greater depth, more steeply inclined soil nails or soil nails oblique to the slope or wall face may be required (Figure 4.5). In such cases, the possible reduction of soil nail capacity should be taken into account in the assessment of the overall stability of the soil nailed structure (see Section 2.2 of Chapter 2 for more details).

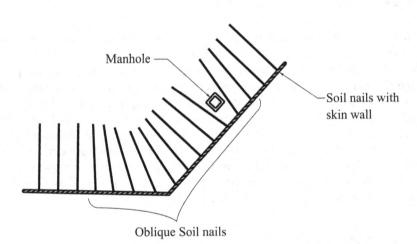

Figure 4.5 Change of nail inclination and layout due to underground obstructions and change of slope geometry.

4.2.6.3 Spacing

The spacing of soil nails depends primarily on the stability requirement of the soil nailed structure. In general, the working load of a soil nail typically ranges from 50 kN to 200 kN. The spacing typically ranges from 0.5 m to 3 m in both horizontal and vertical directions. Since the reinforcing mechanism of soil nailing is to generate a coherent reinforced ground mass, widely spaced soil nails, say greater than 3 m, should be avoided as far as possible. If a wider spacing is required or where the slope face is very steep (e.g. greater than 70°), a tie beam or hard slope facing to connect the nail heads may be warranted to enhance local stability and effective interaction amongst the soil nails.

4.2.6.4 Nail length

Apart from stability consideration, the buildability of the soil nailing scheme in respect of installation method and geology of the site should be taken into account in designing the required lengths of soil nails. For example, ballistic or launched soil nails can normally achieve a length of less than 10 m, whereas driven or "drill-and-grout" soil nails could be up to 20 m or more. Long nail length, say greater than 20 m, should be avoided as far as possible since a greater length may exacerbate construction problems, such as drilling and grouting difficulties. Regarding the length of soil nails, it may be arranged as a uniform pattern with a single nail length for the entire structure, or variable with different nail lengths at different locations of the structure. The benefit of adopting a single nail length for the entire structure is that it can simplify the construction process and facilitate quality control. Figure 4.6 shows the various configurations of nail length. Uniform nail length should be adopted where practicable as possible (Figure 4.6a). In general, variable nail lengths may result in complications during the construction stage. Occasionally, for example, if the deformation of a wall needs to be controlled or the ground condition is highly variable, variable nail lengths may be justified. Past performance of soil nailed walls has indicated that the deformation of a wall can be significantly reduced if the top few rows of soil nails are longer than that required in stability analysis. In contrast, an inverse distribution of nail length (i.e. nail length increases with depth as indicated in Figure 4.6c) should be avoided as far as possible. The use of variable nail lengths should be carefully assessed in the detailed design and quality control plan.

In summary, the preliminary layout and dimensions of soil nails should be determined during the preliminary design stage based on the available information. Actual layout and dimensions of the soil nails should then be assessed and adjusted during the detailed design stage when more site-specific information is available. Table 4.4 summarizes some typical layout and dimensions of soil nailing application which may be used as a quick reference for preliminary design.

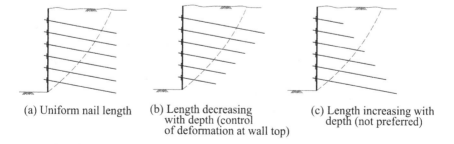

(a) Uniform nail length   (b) Length decreasing with depth (control of deformation at wall top)   (c) Length increasing with depth (not preferred)

Figure 4.6 Different configurations of nail length.

Table 4.4 Typical layout and dimensions of soil nailing application

Design Parameters	Face Angle		
	≤ 45°	> 45° to 70°	> 70°
Pattern	Regular rectangular, square, or triangular		
Inclination	Normal to facing; at an angle of 5° to 20° below horizontal		
Dia. of Drillhole (for "Drill-and-Grout" Nails)	75 mm to 200 mm		
Nail Reinforcement Diameter (Steel)	16 mm to 50 mm		
Nail Reinforcement Length* (L/H)	0.5 to 2.0	0.5 to 2.0	0.5 to 1.2
Nail Spacing	1.5 m to 3.0 m	1.0 m to 2.0 m	0.5 m to 1.0 m
Facing	Soft	Flexible or Hard	Hard

Note: * L and H are the length of soil nails and height of the structure, respectively. In any case, the length of soil nails should preferably be kept within about 20 m to 25 m in length.

4.3 APPROACH FOR DETAILED DESIGN

Soil nailed structures should be designed principally for safety, serviceability, and durability. The uncertainties involved would need to be managed throughout the design process. The sources of uncertainties are primarily attributed to the spatial and temporal variability of geological and hydrogeological profiles, ground strength, loads, methods of analyses, material dimensions and properties, etc. Broadly speaking, there are two approaches for designing a soil nailed structure, namely (i) Allowable Stress Design (ASD) or Working Stress Design (WSD), and (ii) Load and Resistance Factor Design (LRFD), also known as Limit State Design (LSD). These two approaches, i.e. ASD and LRFD, differ in the way that the uncertainties are handled.

For ASD, a single factor of safety is usually employed to compensate for all the uncertainties from both load and resistance. The selection of the required safety factor is essentially based on experience, observed performance, and engineering judgment. ASD has been adopted for many decades and has proven to provide safe and reasonably economical designs, though it may result in an inconsistent level of safety for a structure involving various components with multiple factors of safety. In other words, the reliability level of two soil nailed structures may be different even though they have the same calculated single safety factor (see Section 2.4 in Chapter 2 for more details). The general form of ASD can be conveniently represented by the following expression:

$$\frac{R}{FS} \geq \sum_1^n Q_{di} + \sum_1^n Q_{ti} + \sum_1^n Q_{ei} \tag{4.1}$$

where FS is the factor of safety, R is ultimate resistance, Q_{di}, Q_{ti}, and Q_{ei} for $i = 1$ to n are different types of dead, transient, and extreme loads, respectively.

LRFD, on the other hand, could handle the design uncertainties in a more rational and systematic manner. In the LRFD approach, the uncertainties of load and resistance are reflected by separate parameters, namely load factors and resistance factors. Separate parameters have been used because the nature, variability, and the level of uncertainty associated with load and resistance are different. This results in a more consistent level of safety across different soil nailed structures. The load factors and resistance factors are selected using statistical technique with calibration against the actual load and resistance data such that the joint probability that the actual loads are larger than the design values and the actual resistance is lower than design values would be sufficiently low. In other words, the probability that the resistance of a soil nailed structure is larger than the loads is sufficiently high. The general form of LRFD is represented by the following expression:

$$\frac{R}{\phi} \geq \sum_1^n \gamma_{di} Q_{di} + \sum_1^n \gamma_{ti} Q_{ti} + \sum_1^n \gamma_{ei} Q_{ei} \tag{4.2}$$

where ϕ is resistance factor, R is ultimate resistance and γ_{di}, γ_{ti}, and γ_{ei} for $i = 1$ to n are load factors corresponding to different types of dead loads, Q_{di}, transient loads, Q_{ti}, and extreme loads, Q_{ei}, respectively.

Although the LRFD approach has been adopted by many European countries, the ASD method is still very popular in many parts of the world, such as Hong Kong and Japan. In North America, both ASD and LRFD are being adopted for the design of soil nailed structures. For more details on design standards adopted by different countries for soil nailed structures, reference should be made to Chapter 9.

4.4 DESIGN FOR SAFETY

4.4.1 General

A soil nailed structure becomes unstable when failure or collapse mechanisms develop whereby the applied loads exceed the available resistance provided by either the entire structure or by individual components. Adequate safety margin against potential failure modes must be provided to a soil nailed structure during construction stage and throughout its design life.

The structure should be designed against ultimate or strength limit state corresponding to overall stability as well as stability of individual components, i.e. the limit state is reached if any one of its safety performance criteria is satisfied.

4.4.2 Models

Upon the identification of all potential failure modes in the preliminary design stage, one should formulate appropriate design models to reflect the heterogeneity of ground conditions as well as the loading conditions, such as groundwater levels, that need to be designed for during the lifetime of the structure. There should be adequate engineering geological input to the ground investigation and formulation of representative ground and groundwater models for stability assessment and design verification during construction.

In order to ensure the adequacy of engineering geological input in the process of design, a three-step approach comprising "geological," "ground," and "design" models may be adopted. As the first step, a geological model is used to characterize a site where the focus is placed on geological, geomorphological, and hydrogeological features, and characteristics that are relevant to an engineering project. A ground model then builds on the geological model and integrates the range of engineering parameters and ground conditions that need to be considered in the design. It refines the geological model by defining and characterizing bodies of ground with similar engineering properties, and identifies boundaries at which changes in geotechnical conditions may occur. A design model, on the other hand, is concerned primarily with assessment of ground response to the proposed works for use in geotechnical assessment or engineering design. Design models for empirical, prescriptive, and quantitative designs depend on the engineering application, degree of conservatism in the empirical/prescriptive design, and the level of geotechnical risk. Figure 4.7 illustrates the roles of engineering geology, soil mechanics, and rock mechanics within the geo-engineering practice.

4.4.3 Safety margin

In general, the risk concept should be adopted in defining the safety margin and degree of rigor required for the design of a soil nailed structure. The fundamental philosophy of a risk-based design approach is to assess the probability of failure of the structure (hazard) and the potential consequence upon its failure, so as to arrive at an acceptable or tolerable safety level through appropriate risk management measures as necessary. For example, a 3 × 3 risk matrix is proposed in the UK for the design of a soil nailed structure (Table 4.5) in which the risk is categorized as ranging from "negligible risk" to "unacceptable risk." Depending on the risk level,

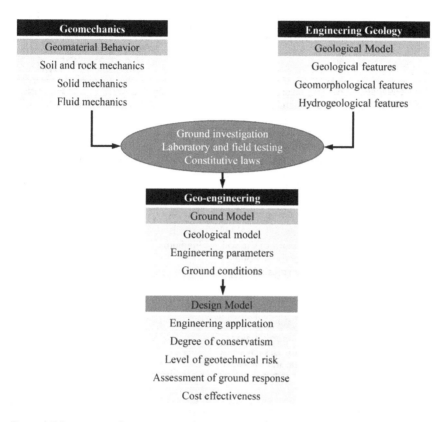

Figure 4.7 International perspective of the roles of engineering geology, soil mechanics, and rock mechanics within geo-engineering practice. (Modified Bock, 2006.)

the degree of rigor required for the entire design process encompassing site investigation, detailed design, material selection, construction method, and site supervision during construction would be different.

4.4.3.1 External stability (ASD approach)

Different parts of the world may adopt different approaches in defining the required safety margin. Some consider explicitly both the probability of failure (hazard) and consequence of failure, and others do not. For example, in Hong Kong the consequences of failure with respect to life and economic loss are explicitly considered as illustrated in Tables 4.6 and 4.7, respectively. Depending on the consequence-to-life and the economic consequence categories, the required safety margin in terms of factor of safety, F_G, for external stability of a newly formed cut slope and an existing cut slope reinforced by soil nails can be defined according to Tables 4.8 and 4.9,

Table 4.5 Risk matrix proposed for design of a soil nailed structure (Modified Phear et al., 2005, BSI, 2013)

		Probability of Failure		
		Low	*Medium*	*High*
Consequence of Failure	Low (e.g. failure of the soil nailed structure would result in minimal damage to infrastructures or disruption to services)	Negligible risk	Low risk	Medium risk
	Medium (e.g. failure of the soil nailed structure would result in limited damage to infrastructures or moderate disruption to services)	Low risk	Medium risk	High risk
	High (e.g. failure of the soil nailed structures would result in serious injury or fatalities, significant damage to infrastructures or disruption to services; or the soil nailed structure directly supports major infrastructures, such as trunk roads and railways)	Medium risk	High risk	Unacceptable risk

respectively. One may note that the probability of failure of the structure has been implicitly considered in the required minimum factor of safety.

In the United States, the recommended safety margin for external stability of soil nailed structures using ASD approach is given in Table 4.10. As some areas in the United States may be prone to earthquake, it is also required to consider extreme seismic loads in assessing the external stability of the structure, though the corresponding required minimum factors of safety are lower than those for nominal loads.

Indeed, the required minimum factor of safety for a soil nailed structure varies from place to place. The setting of this safety value depends on the risk level of the structure, the level of uncertainty involved in the design and construction process, and the risk that society is prepared to tolerate or accept.

4.4.3.2 Internal stability (ASD approach)

In Hong Kong, the recommended minimum factors of safety against the four common modes of internal failure of a soil nail, viz. (i) tensile failure of soil nail reinforcement, (ii) pullout failure at soil-grout interface, (iii) pullout failure at grout-reinforcement interface, and (iv) bearing failure of soil underneath nail head, are given in Table 4.11. This is independent of the failure

Table 4.6 Typical examples of slope failures in each consequence-to-life category (GEO, 2008)

Examples	Consequence-to-life		
	Category 1	*Category 2*	*Category 3*
(1) Failures affecting occupied buildings (e.g. residential, educational, commercial, or industrial buildings, bus shelters, railway platforms).	✓		
(2) Failures affecting buildings storing dangerous goods.	✓		
(3) Failures affecting heavily used open spaces and recreational facilities (e.g. sitting-out areas, playgrounds, car parks).		✓	
(4) Failures affecting roads with high vehicular or pedestrian traffic density.		✓	
(5) Failures affecting public waiting areas (e.g. bus stops, petrol stations).		✓	
(6) Failures affecting country parks and lightly used open-air recreational areas.			✓
(7) Failures affecting roads with low traffic density.			✓
(8) Failures affecting storage compounds (non-dangerous goods).			✓

consequence of the slope. Also, as discussed in Chapter 2, given that the contributing effect of the shear and bending resistance to the safety of soil nailed structures is small as compared with the tensile resistance, the shear and bending strength contributions from soil nails are conservatively disregarded.

In the United States, as the majority of soil nails are applied to the retaining wall, designers are required to consider internal failure modes in relation to (i) the tensile failure of soil nail reinforcement, (ii) the pullout failure at the interface between soil and grout, and (iii) the flexural and punching shear failure of wall facing. The corresponding recommended minimum safety factors are given in Table 4.12. Unlike the practice in Hong Kong, there is no requirement for checking the pullout failure at the grout-reinforcement interface as past experience in the United States suggests that this is not a critical failure mode given threaded bars and relatively high strength grout that have long been used in practice. Also, the contribution of shear and bending capacity of soil nails could be conservatively disregarded and hence the checking of this failure mechanism is not required.

4.4.3.3 Partial safety factors (LRFD approach)

Apart from using Allowable Stress Design (ASD) to assess the external and internal failure mechanisms, one may adopt Load and Resistance Factor

Table 4.7 Typical examples of slope failures in each economic consequence category (GEO, 2008)

Examples	Economic Consequence		
	Category A	Category B	Category C
(1) Failures affecting buildings, which could cause excessive structural damage.	✓		
(2) Failures affecting essential services, which could cause loss of that service for an extended period.	✓		
(3) Failures affecting rural or urban trunk roads or roads of strategic importance.	✓		
(4) Failures affecting essential services, which could cause loss of that service for a short period.		✓	
(5) Failures affecting rural (A) or primary distributor roads, which are not sole accesses.		✓	
(6) Failures affecting open-air car parks.			✓
(7) Failures affecting rural (B), feeder, district distributor, and local distributor roads, which are not sole accesses.			✓
(8) Failures affecting country parks.			✓

Design (LRFD), also known as Limit State Design (LSD), to design a soil nailed structure. In this approach, a set of partial factors is applied to load and resistance in order to minimize the risk of the soil nailed structure attaining different potential failure mechanisms. In general, the design value of an action should be obtained by multiplying the representative value by the load partial factor, whereas the design value of the soil nail resistance including material properties should be obtained by dividing the characteristic values of material properties and soil nail resistance by appropriate partial factors. For illustration purposes, the sets of partial factors

Table 4.8 Recommended minimum factor of safety, F_G, against external failure for new soil nailed cut slopes in Hong Kong (GEO, 2008)

Economic Consequence	Consequence-to-life		
	Category 1	Category 2	Category 3
Category A	1.4	1.4	1.4
Category B	1.4	1.2	1.2
Category C	1.4	1.2	> 1.0

Notes: (1) In addition to a minimum factor of safety of 1.4 for a 10-year return period rainfall, a slope in the consequence-to-life category 1 should have a factor of safety of at least 1.1 for the predicted worst groundwater conditions.

(2) The factors of safety given in this table are recommended minimum values. Higher factors of safety might be warranted in particular situations in respect of loss of life and economic loss.

Table 4.9 Recommended minimum factor of safety, F_G, against external failure for existing cut slopes upgraded by soil nails in Hong Kong (GEO, 2008)

Consequence-to-life	Category 1	Category 2	Category 3
Minimum Factor of Safety	1.2	1.1	> 1.0

Notes: (1) These factors of safety are appropriate only where rigorous geological and geotechnical studies have been carried out (which should include a thorough examination of maintenance history, groundwater records, rainfall records, and any monitoring records), where the slope has been standing for a considerable time without instability or distress, and where the loading conditions, the groundwater regime, and the basic form of the modified slope remain substantially the same as those of the existing slope. Otherwise, the standards specified for new slopes given in Table 4.8 should be adopted.

(2) The factors of safety given in this table are recommended minimum values. Higher factors of safety might be warranted in particular situations in respect of loss of life and economic loss.

Table 4.10 Recommended minimum factor of safety for design of soil nailed walls in the United States (Lazarte et al., 2003 & 2015)

Failure Mode	Permanence	Minimum Factor of Safety	
		Static Loads	Seismic Loads
Global or Overall	Permanent	$F_G = 1.5$	$F_G = 1.1$
	Temporary	$F_G = 1.25$ to 1.33	NA
Sliding	Permanent	$F_S = 1.5$	$F_S = 1.1$
	Temporary	$F_S = 1.3$	$F_S = 1.1$
Bearing	Permanent	$F_B = 3.0$	$F_B = 2.3$
	Temporary	$F_B = 2.5$	$F_B = 2.3$

Table 4.11 Recommended minimum factor of safety against internal failure of a soil nail in Hong Kong (GEO, 2008)

Mode of Internal Failure	Minimum Factor of Safety
Tensile failure of soil nail reinforcement	$F_T = 1.5$
Pullout failure at soil-grout interface	$F_P = 1.5$ [Note 1]
	$F_P = 2.0$ [Note 2]
Pullout failure at grout-reinforcement interface	$F_R = 2.0$
Bearing failure of soil underneath nail head	$F_B = 3.0$

Notes: (1) For soil nails carrying transient loads and bonded in weathered granite or volcanic rocks.

(2) For soil nails carrying sustained loads or for soil nails carrying transient loads and bonded in soils other than weathered granite or volcanic rocks.

currently adopted in the UK are given in Table 4.13. In UK practice, there are two sets of partial factors and designers are required to ensure that the ultimate limit state corresponding to the pertinent external and internal failure modes will not occur with either set. Also, the partial factors given in Table 4.13 do not explicitly take account of whether the structure is of a temporary or permanent nature. In practice, this nature should be reflected in assessing the characteristic design values.

Table 4.12 Recommended minimum factor of safety against internal failure of a soil nail in the United States (Lazarte et al., 2003 & 2015)

Mode of Internal Failure	Permanence	Minimum Factor of Safety	
		Static Loads	Seismic Loads
Tensile failure of soil nail reinforcement	Permanent	$F_T = 1.8$	$F_T = 1.35$
	Temporary		
Pullout failure at soil-grout interface	Permanent	$F_P = 2.0$	$F_P = 1.5$
	Temporary		
Flexural failure of wall facing	Permanent	$F_{FF} = 1.5$	$F_{FF} = 1.1$
	Temporary	$F_{FF} = 1.35$	$F_{FF} = 1.1$
Punching failure of wall facing	Permanent	$F_{PF} = 1.5$	$F_{PF} = 1.1$
	Temporary	$F_{PF} = 1.35$	$F_{PF} = 1.1$

Table 4.13 Partial factors for soil nail design in the UK (Modified BSI, 2013)

	Description of Load/Resistance		Partial Factors	
			Set 1	Set 2
Actions	Self-weight of soil, W	Destabilizing	$\gamma_g = 1.35$	$\gamma_g = 1.0$
		Stabilizing	$\gamma_g = 1.0$	$\gamma_g = 1.0$
	Permanent surcharge, q_p	Destabilizing	$\gamma_{qp} = 1.35$	$\gamma_{qp} = 1.0$
		Stabilizing	$\gamma_{qp} = 1.0$	$\gamma_{qp} = 1.0$
	Variable surcharge, q_v	Destabilizing	$\gamma_{qv} = 1.5$	$\gamma_{qv} = 1.3$
		Stabilizing	$\gamma_{qv} = 0$	$\gamma_{qv} = 0$
	Groundwater pressure, u	Destabilizing	$\gamma_u = 1.0$	$\gamma_u = 1.0$
		Stabilizing	$\gamma_u = 1.0$	$\gamma_u = 1.0$
Material properties	Angle of shearing resistance, $\tan \phi'$		$\gamma_{\tan \phi'} = 1.0$	$\gamma_{\tan \phi'} = 1.3$
	Effective cohesion, c'		$\gamma_{c'} = 1.0$	$\gamma_{c'} = 1.3$
	Undrained shear strength, c_u		$\gamma_{cu} = 1.0$	$\gamma_{cu} = 1.4$
	Unit weight, γ		$\gamma_\gamma = 1.0$	$\gamma_\gamma = 1.0$
Soil Nail Resistances	Bond stress, τ_b	Empirical	$\gamma_{\tau b} = 1.1$	$\gamma_{\tau b} = 1.5$
		Effective Stress	$\gamma_{\tau b} = 1.1$	$\gamma_{\tau b} = 1.5$
		Total Stress	$\gamma_{\tau b} = 1.1$	$\gamma_{\tau b} = 1.5$
		Pullout Test	$\gamma_{\tau b} = 1.1-1.7$	$\gamma_{\tau b} = 1.5-2.25$
	Tendon strength, T		$\gamma_s = 1.0$	$\gamma_s = 1.15$ for steel

4.4.4 Methods of analysis

Different analytical methods are available for assessing the stability of a soil nailed structure. The majority of these comprise conventional methods of slope stability analysis including two-dimensional limit equilibrium analysis based on the method of slices, where the sliding soil mass is modeled as a rigid body divided into a number of slices (e.g. Janbu, 1954; Bishop, 1955; Morgenstern & Price, 1965; Spencer, 1967). As with traditional slope stability analyses, potential failure surfaces of different shapes, circular or non-circular, are evaluated until the most critical failure surface is obtained, i.e. the one corresponding to the lowest factor of safety. Nevertheless, numerical comparisons among different methods show that differences in the geometry of the failure surface do not result in a significant difference in the calculated factors of safety (e.g. Long et al., 1990). However, in choosing the method of limit equilibrium analysis, designers should consider whether the method satisfies all the conditions of equilibrium. The calculated factors of safety derived from methods that consider force equilibrium or moment equilibrium only may not give correct results. Therefore, methods that satisfy both force and moment equilibrium should be used for analysis (Shiu et al., 2007). Furthermore, commercially available general slope stability programs are more commonly used to evaluate the overall stability of a soil nailed structure than those dedicated for soil nail design. Given this context, it is important that designers should carefully consider how to realistically incorporate the stabilizing forces provided by soil nails into the analysis.

In case commercial slope stability software packages are not readily available, one may carry out limit equilibrium analyses based on simplified methods such as the single, two-part, or three-part wedge analyses. These wedge methods provide simple procedures for obtaining a relatively conservative solution by hand calculations and are particularly suitable for soil nailing design. Figure 4.8 shows a typical geometry of the two-wedge mechanism of a slope. The constraints on the method are that the inter-wedge boundary should be vertical, and that the lower wedge should intersect the toe of the slope. As a good practice, designers may use computer programs to analyze and design soil nailed structures, and carry out wedge analysis as an independent verification of the computer's solution. More details on the two-wedge method can be found in the DOT Advice Note, HA 64/94, published by the Highway Agency of the Department of Transport in the UK (DOT, 1994).

Although limit equilibrium analyses have long been used and proven to be a convenient means for assessing the stability of soil nailed structures, a limitation of these methods is that they do not provide a prediction of deformations, nor do they consider the deformation required to mobilize the resisting forces in the soil nails. These methods cannot provide a thorough description of the contribution of each soil nail to global stability based on the deformation pattern of the ground mass. Nevertheless, past experience indicates that these methods tend to provide safe and economical solutions in most cases.

For certain projects, a stress-strain analysis may be required for assessing the anticipated performance of a soil nailed structure with respect to the capacity of soil nails that have to deviate from optimum inclinations

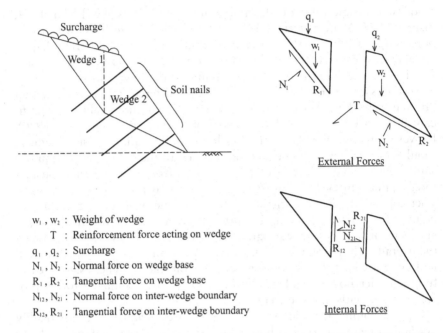

w$_1$, w$_2$: Weight of wedge

T : Reinforcement force acting on wedge

q$_1$, q$_2$: Surcharge

N$_1$, N$_2$: Normal force on wedge base

R$_1$, R$_2$: Tangential force on wedge base

N$_{12}$, N$_{21}$: Normal force on inter-wedge boundary

R$_{12}$, R$_{21}$: Tangential force on inter-wedge boundary

Figure 4.8 **Typical geometry of the two-wedge mechanism of a slope. (Modified DOT, 1994.)**

because of site constraints, or for ground deformation assessment. For instance, if the soil nails are steeply inclined, the tensile forces that can be mobilized in the soil nails will be less than those of gently inclined soil nails. In this case, numerical methods (e.g. finite element or finite difference methods) may be used for the analysis. Examples can be found in Shiu & Chang (2006) and Singh & Babu (2010). There are different algorithms that can be adopted, such as the strength reduction method and the approach of coupling numerical analysis with the limit equilibrium method (Krahn, 2003). In selecting a numerical method for deformation analysis, the following should be considered:

- Ground and groundwater conditions.
- Sensitivity of the numerical model to changes of geometry during construction.
- Previous or comparable experience.
- Calibration of the numerical method or constitutive model being adopted in the analysis.

In the context of ground deformation, horizontal deformation analyses are fraught with difficulties. In many cases, they can only be taken as approximate or else it is simply assumed that the usual factors of safety against external or internal stability failure will ensure that deformations will be within tolerable limits based essentially on past experience. Vertical deformation analyses are obtained from conventional settlement computations, with particular emphasis on differential settlements, both longitudinally along the wall face, and transversely from the face to the end of the reinforced soil volume. The results may affect the choice of facing or facing connections.

4.5 DESIGN FOR SERVICEABILITY

4.5.1 General

Apart from stability, the performance of a soil nailed structure should satisfy serviceability requirements, which are defined during the preliminary design stage. Serviceability limit state refers to the conditions that normally do not involve collapse of the soil nailed structure, but it may impair normal operation and its function. Common serviceability limits include horizontal deformation, bulging, differential settlement, facing distortion, etc. Although failure to meet the serviceability requirements may not lead to collapse, the normal and safe operation of the structure could be impaired. The following are some typical scenarios that do not meet the serviceability requirements:

- Excessive wall deformation resulting in adverse visual appearance.
- Cracking of hard facing.
- Excessive bulging of flexible or soft facing.
- Excessive total or differential ground settlement leading to malfunction of surface or underground utilities or services, or affecting nearby structures.

4.5.2 Deformation

As the soil nailed structure comprises a passive reinforcement system, some deformation of the structure is expected during construction as well as throughout its service life. Figure 4.9 shows schematically a soil nailed structure that deforms outward due to incremental rotation about the toe of the structure during construction. Most of the deformation should occur during or shortly after excavation. Post-construction deformation is generally small unless there is deterioration of the structure, its foundation, or the retained soil. Creep of retained fine-grained soils may take place with the consequent increase of tensile forces in the soil nails and deflection of the structure after construction. According to Lazarte et al. (2015), post-construction deformations may increase up to 15% beyond those observed soon after construction. Fine-grained soils of high-plasticity (Plasticity

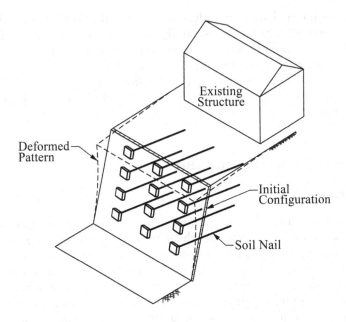

Figure 4.9 Deformation of a soil nailed structure.

Index, PI > 20) and high water contents (Liquidity Index, LI > 0.2) tend to deform for a long period of time due to their potential for creep. In general, the maximum horizontal deformation normally occurs near the top of the structure and decreases gradually toward the toe of the structure. The vertical deformation (i.e. settlement) is in the same order of horizontal deformation.

The deformation of a soil nailed structure is governed by various factors, which include the ground profile, soil stiffness, groundwater condition, layout, orientation and inclination of soil nails, type of slope facing, construction sequence, and workmanship. In estimating deformation of a new excavation, considerations should be given to the following factors:

- Self-weight, in particular if top-down construction method is adopted in the case of a new excavation.
- Construction sequence and rate.
- Excavation height for each lift.
- Compressibility of the ground.
- External forces, such as water pressures, surcharge, and impact loads.
- Seismic loads.
- Prolonged swelling or heave in cohesive soils.

Due to the complexity of interaction between soil nails and the ground, precise prediction of ground deformation is difficult. In practice, the

Table 4.14 Preliminary estimation of wall displacements (Clouterre, 1991)

	Weathered Rocks/Stiff Soils	Sandy Soils	Clayey Soils
$\delta_v = \delta_h$	H/1000	2H/1000	4H/1000
κ	0.8	1.25	1.5

Note: δ_v is the vertical displacement, δ_h is the horizontal displacement, H is the height of the wall, κ is the coefficient for estimation of the influence zone behind the wall face where the ground deformation is negligible, $\delta_0 = \kappa \, [1-\tan(90°-\beta_s)]H, \beta_s$ is the angle of the wall.

ground deformation can be estimated based on empirical correlation, relevant preceding case studies, or numerical analysis, or a combination of the three.

Under normal circumstances, limit equilibrium methods may be adopted for design of a soil nailed structure. For the purpose of a preliminary check, empirical approaches may be adopted to estimate the likely vertical and horizontal deformations. For example, Clouterre (1991) suggests that the maximum horizontal and vertical displacements at the top of a soil nailed wall in a new excavation can be estimated based on soil type and height of the wall (Table 4.14).

Relevant case studies may also help to predict the likely ground deformation of a soil nailed structure. Readers are recommended to refer to Table 2.4 of Chapter 2 in which information on deformation performance of soil nailed excavations in different places with heights ranging from 7 m to 50 m during the course of construction has been collected. However, similarity of the design and the preceding reference cases should be validated in respect of the ground profile and geometry of soil nails, ground and groundwater conditions, construction method and sequence, loading conditions, etc.

Where deformation of a soil nailed structure is a cause for concern (e.g. need for tight control due to presence of sensitive receivers or services in the vicinity) or where a more accurate estimation is warranted, detailed deformation analysis should be carried out. For example, numerical modeling using finite element or finite difference computer programs, together with realistic constitutive models, should be used for the analysis. Nonetheless, designers should be aware of the factors that may affect the accuracy of the deformation calculation. These include the accuracy of constitutive soil/rock model, interaction between the soil nails and the ground, etc. Calibration with past local cases supplemented with construction monitoring may help to improve the accuracy of the prediction. Necessary mitigation measures, such as by using battered wall/slope, installing longer soil nails at the top of excavation, increasing the ratio of nails per unit area of wall/slope, and by using temporary pre-stressed ground anchors in conjunction with soil nails, may be employed to limit the amount of deformation.

4.6 DESIGN FOR DURABILITY

For most applications, a permanent soil nailed structure is designed to have a service life of about 50 to 120 years. As soil nails are commonly installed in ground subject to seasonal variation in environmental conditions, such as water content, humidity, temperature, and the degree of exposure to environment, the durability of soil nails and the associated components would inevitably be influenced by these factors. Soil nailed structures must be sufficiently durable, so that they are capable of withstanding attack from the existing as well as envisaged corrosive environment without unduly affecting their stability and serviceability during the service life.

Steel bars are the most common type of soil nail reinforcement. Steel reacts with water and oxygen to form oxides and hydroxides. In the context of soil nailing, differential aeration cells would be established in the nail reinforcement when it passes through the ground with different permeabilities. For the segments of nail reinforcement that are surrounded by soil of high concentration of dissolved oxygen, cathodic areas would be established. In contrast, anodic areas would be formed at those segments surrounded with soil of relatively low oxygen concentration. Corrosion would then occur near the boundaries of the cathodic areas. Corrosion can be broadly classified into the following three categories:

- Generalized attack across the exposed surface resulting in uniform corrosion.
- Localized attack to particular exposed surface resulting in pitting corrosion.
- Cracking as a result of either hydrogen embrittlement or environmentally assisted stress corrosion cracking due to conjoint action of internal and external static tensile stress and localized corrosion.

In considering the durability of a soil nailed structure, a designer should understand and consider the possible environmental conditions that the soil nails would be exposed to during the service life. It is easier to determine with reasonably accuracy the environmental conditions that exist at the time of construction by carrying out site investigation compared to those during the service life of the soil nailed structure. A detailed assessment of ground aggressivity can be made during site investigation by means of physical and chemical testing in the laboratory, review of site records, and field observations.

Nonetheless, it is not uncommon for field or laboratory tests for assessing ground aggressivity not to be included in the site investigation, unless soil nailing has been identified as a possible design option at an early stage without the necessary data. This renders it difficult for designers to assess the corrosion potential of the ground. As a contingency measure, one may allow sufficient time in the project program such that supplementary site investigation, if needed, could still be carried out to assess the corrosion potential of the site during construction stage.

Sometimes, the aggressivity of a site may still be assessed, without carrying out site investigation, from the site setting, development history, and the nature and extent of facilities that may affect the site. For example, a site should be regarded as having high corrosive potential if it has been, or is likely to be, affected by leakage or discharge from old development, sewage treatment facilities, industrial facilities, or livestock facilities, or it is in the vicinity of sources of stray currents like electricity substation, electrified rail system, or tramway system. In contrast, a site may be regarded as non-aggressive if it shows no signs of seepage or high groundwater levels that could bring corrosive agents from distant sources into contact with the soil nails.

Upon establishment of the aggressivity of the site, appropriate corrosion protection measures should be provided to the soil nails and their accessories with due consideration given to the environmental conditions that the soil nailed structure will be exposed to during the service life. Common corrosion protection measures to steel reinforcement and components include zinc coating (i.e. galvanization), epoxy coating, sacrificial steel, and encapsulation. In determining the types of corrosion protection measures, one should consider pertinent factors like the reliability and long-term performance of the measures, potential adverse effect on bond strength at the grout-reinforcement interface, cost, availability in the market, ease of handling, and quality control on site. Apart from providing the necessary corrosion protection measures to steel reinforcement, one may select soil nail reinforcement made of corrosion-resistant materials, such as fiber reinforced plastics (FRP) (see also Section 4.10.1). Chapter 5 gives more details on the durability of soil nails.

4.7 DRAINAGE

4.7.1 General

Water, in the form of surface water flows, groundwater inflows, or building up of pore water pressures within or behind the reinforced soil mass, is one of the major hazards that can affect the performance of a soil nailed structure during construction as well as throughout its service life. Surface runoff and groundwater should be properly controlled to ensure satisfactory performance of a soil nailed structure, both during construction and service life. Surface and subsurface drainage should be provided to avoid the build-up of excessive water pressure within and behind the soil nailed structure as well as to limit the detrimental effect that surface water or groundwater may have on the structure. The drainage system for a soil nailed structure can be broadly classified into three categories, namely (i)

Table 4.15 Category of drainage system for a soil nailed structure

Category of Drainage System	Purposes of Application
Surface Drainage	To prevent uncontrolled surface runoff, mainly arising from rainfall or snowmelt, from infiltrating into the reinforced zone or flowing over a soil nailed structure edge. The drainage system is commonly in the form of a network of surface channels or ditches.
Facing Drainage	To prevent the accumulation of water behind the facing of a soil nailed structure. It can be in the form of short drainage pipes (i.e. weep holes), a geosynthetic filter behind the facing, or use of permeable no-fines concrete as the facing.
Subsurface Drainage	To prevent the build-up of excessive water pressure within and behind the soil nailed structure. The drainage is usually in the form of slotted tubes with impermeable invert, known as horizontal drains or raking drains, that are generally longer than the soil nail reinforcement, so as to control the development of excessive water pressure within and behind the reinforced zone. The density of this drainage depends on the ground and groundwater conditions. As a rule of thumb, the spacing is normally similar to that of soil nails.

surface drainage, (ii) facing drainage, and (iii) subsurface drainage. The density and type of drainage to be provided depend on the geometry of the soil nailed structure, surface and subsurface water conditions, and geology of the site. Longevity and the level of maintenance required during the service life of a soil nailed structure should also be considered in the design. As a general design principle, a drainage system should be of high redundancy and be capable of self-cleansing. Table 4.15 shows the application objectives of the three categories of drainage systems. Some typical types of drainage measures that may be adopted for a soil nailed structure are summarized in Figure 4.10.

In general, a detailed appraisal of the likely flow paths of surface runoff and the potential for concentration of surface water flow affecting a soil nailed structure should be undertaken in the design of the surface drainage system. Sufficient redundancy should be allowed for in the design based on engineering judgment, with due regard given to the site environmental setting and consequence in the event of blockage of the surface drainage channels. Appropriate facing and subsurface drainage measures are also essential to prevent the accumulation of water behind the facing and development of high base or perched groundwater levels. Leakage from underground water-carrying services can adversely affect the performance of a soil nailed structure. Where such services are present within or in the vicinity of the reinforced zone, action should be taken to reduce the risk of leakage.

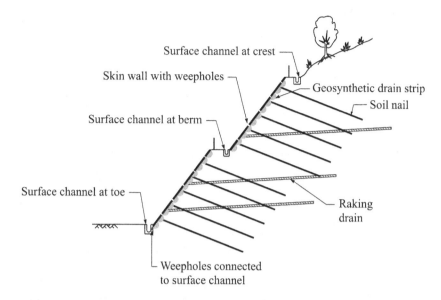

Surface channel at crest

Skin wall with weepholes

Surface channel at berm

Surface channel at toe

Geosynthetic drain strip

Soil nail

Raking drain

Weepholes connected to surface channel

Figure 4.10 Common types of drainage measures for a soil nailed structure.

4.7.2 Temporary drainage system during construction

During construction, sufficient temporary drainage for controlling surface water runoff and subsurface flow associated with either perched water or localized seepage areas should be provided at all times, especially during the rainy season, in order to avoid any adverse effects of uncontrolled concentrated surface or subsurface water flow. Common drainage measures include surface water interceptor ditches or drainage channels, temporary dewatering through subsurface drains, or well points installed beyond the length of the soil nails, etc. Local high groundwater level may be encountered during construction and it can cause significant problems to the drilling and grouting operations. Temporary horizontal drains may be installed to lower the water level to facilitate the soil nail installation works.

The temporary site drainage should be regularly maintained and cleared of any blockage to ensure that the drains remain functional during heavy rainfall. Where appropriate, part of the permanent drainage measures, e.g. crest drain and the associated discharge points, should be constructed at an early stage of the works to enhance the temporary drainage provision. During the construction of subsurface drains, due attention should be paid to avoiding damage to the installed soil nails adjacent to the drains.

4.7.3 Permanent drainage system

4.7.3.1 Surface water control

Surface water control measures include installing an interception surface channel or ditch at the toe and crest of a soil nailed structure to prevent surface runoff from infiltrating into the reinforced zone or flowing over the edge of the structure. A vegetation or hard cover, such as shotcrete protective facing, may also be used to reduce or retard water infiltration into the ground. Potential convergent surface water flow should be diverted away from the structure as far as possible or should be directed downslope, preferably with no or little change in flow direction, by means of drains with adequate capacity. Consideration should be given to providing an upstand for the crest drainage channel of the structure in order to minimize possible uncontrolled spillage of surface water, and increasing the channel gradient and size. Special attention should be given to the layout and detailing of the surface drainage system to ensure adequate flow capacity and containment of flow within the channels, together with adequate discharge capacity at the downstream side. For instance, abrupt changes in flow directions, which can be conducive to over-spilling or overflow along the channels, should be avoided. Environmental factors, such as potential sources of concentrated surface water flow which may adversely affect the stability of the structure, should be dealt with properly. Junctions of surface drainage channels should be properly detailed to avoid excessive turbulence and splashing. A smooth transition of alignment should be provided at the junction of berm channels and downslope channels where practicable in order to improve the hydraulics of surface water flow. Baffle walls, or catchpits, should be provided at junctions of channels, if deemed necessary, to minimize over-spilling or overflow. Baffle walls may be preferred as catchpits could be susceptible to blockage (e.g. by erosion debris or dead vegetation).

4.7.3.2 Groundwater control

Where the build-up of groundwater pressure during severe rainfall is likely to be so rapid that drainage from the surface of a soil nailed structure alone may not be adequate to avoid failure, subsurface drainage provisions as discussed in the following should be provided:

Raking drains. Raking drains, also known as horizontal drains, can effectively lower the groundwater level and relieve the groundwater pressure at depth. The main advantages of raking drains are that they are relatively simple and quick to install and they rely on gravity drainage. Raking drains are used extensively worldwide. For example, in Hong Kong, raking drains are commonly in the form of polyvinyl chloride (PVC) slotted or perforated tubes with impermeable invert, typically 50 mm to 75 mm in diameter, inclined upward at 5° to 10° above the horizontal to facilitate gravitational flow of water. Depending on the groundwater conditions, raking drains

Table 4.16 Application of raking drains for different purposes

Location of Raking Drains	Purposes of Application
Upper Part of a Soil Nailed Structure	To control the development of a potential perched water table in a more permeable soil stratum overlying a less permeable stratum.
Lower Part of a Soil Nailed Structure	To control the transient rise of the main groundwater level during or following rainfall.
Specific Areas	To facilitate drainage and relieve the water pressures at specific locations where persistent seepage or preferential flow paths are present.

are usually designed to be in staggered rows with spacing varying typically from 3 m to 5 m. It is good practice to incorporate a fabric filter into the drain to prevent erosion or clogging. Raking drains are typically longer than the length of the soil nails and serve to intercept any groundwater behind the soil nailed zone. Raking drains are commonly installed after all the soil nails at elevations higher than the drains have been grouted to prevent potential intrusion of cement grout into the slotted pipes. The drains typically exit through the face of the structure. Depending on the application, raking drains can be installed at different locations of the soil nailed structure as suggested in Table 4.16.

Toe drains. For soil nailed structures affected by high groundwater level, construction of toe drains provides an effective means of lowering the groundwater level close to the lower part of the structure. Toe drains also provide a proper drainage discharge point to other drainage measures, such as facing drainage. Toe drains are usually in the form of partially or fully perforated PVC pipes buried in compacted fill or no-fines concrete.

Counterfort drains. For soil nailed structures that are liable to experience a rapid build-up of groundwater pressure, such as development of a perched water table in a relatively thin surface mantle of loose material like colluvium or fill overlying weathered rock, the use of raking drains alone may not necessarily be able to provide sufficient drainage capacity to quickly relieve the transient groundwater pressure and avoid failure. In such cases, counterfort drains could be used, either on their own or in combination with raking drains. Counterfort drains should be extended into the underlying less permeable ground. If this cannot be achieved, raking drains should be provided to intercept any groundwater flow beneath the counterfort drains. Particular care should be taken to ensure that water tightness of the "impermeable" membrane at the base of the drain is achieved during construction.

4.7.3.3 Control of water affecting the facing

Shallow drains (weep holes). It is good practice to provide shallow drainage, also known as weep holes, for impermeable surface cover, so as to provide drainage for shallow or perched groundwater occurring close to the facing, in particular at locations where localized seepages are encountered or anticipated. Weep holes usually comprise PVC pipes of 50 mm to 100 mm in diameter, and with lengths varying from 300 mm to 500 mm. Under special circumstances, the length of weep holes may extend to 1 m long. In general, weep holes are installed at a density of approximately one drain per 10 m² of the facing. Weep holes are also used as the terminating point of vertical strip drains to allow any collected water to pass through the wall. They are typically installed after nail construction to prevent potential intrusion of cement grout into the pipes.

Geosynthetic drain strips. Inadequate drainage behind hard surface cover (e.g. shotcrete) can be a contributory cause of failure on soil nailed structures with subsurface seepage flow. Geosynthetic composite drainage material may be installed behind the hard surface cover, together with the provision of no-fines concrete toe or relief drains in order to minimize the build-up of groundwater pressure. This is particularly important at locations where preferential flow paths, such as soil pipes, erosion channels, or holes left behind by rotten tree roots or burrowing animals, exist in the ground behind the hard surface cover. These drainage elements comprise strips of synthetic material approximately 300 mm to 400 mm wide. They are placed in vertical strips against the slope or excavation face along the entire depth of the soil nailed structure. It is important to avoid sliding failure at the interface between the soil and geosynthetic composite drainage material by providing proper anchorage, and ensuring that there are no significant gaps at the interface which may result in erosion.

It should be noted that geosynthetic composite drainage material has limited drainage capacity. However, it is likely to be suitable for relieving groundwater pressure in the soil close to the hard surface cover. Where a larger drainage capacity is required, the use of a no-fines concrete cover should be considered. Installation of geosynthetic composite drainage material may be difficult if the slope surface is irregular, e.g. within landslide scars. In such cases, the use of no-fines concrete may be more convenient. Also, a designer should explicitly provide the construction and inspection guidance for this type of drainage measure, so as to ensure good performance of the measure.

No-fines concrete cover. No-fines concrete has good drainage capacity and its dead weight offers some stabilization effect. It can be built conveniently against an irregular ground profile to give a uniform surface, and if used properly in conjunction with a geotextile filter or geosynthetic composite drainage material, it can be very effective in controlling surface instability and erosion. Loose material on the surface should be removed before placing the no-fines concrete. Care should be exercised during placement of no-fines concrete to avoid damaging and blocking the geotextile filter or geosynthetic composite drainage material. If necessary, an additional protective layer of geotextile filter or sand bags may be placed over the geotextile filter or geosynthetic composite drainage material to protect it from damage during placement of the no-fines concrete.

The no-fines concrete cover should be founded on firm ground to ensure stability. Benching of the concrete into the slope should be considered to improve stability, especially with steep slopes or excavations. Galvanized steel dowel bars may also be used to tie the no-fines concrete block and geotextile filter (or geosynthetic composite drainage material) to the soil nailed structure.

Relief drains. Where there are signs of potential seepage sources (e.g. rock joints or preferential flow paths with signs of seepage) behind a hard surface cover, relief drains should be provided. It is important to avoid sliding instability at the interface between the competent materials and the geosynthetic material by providing proper anchorage, such as nails and plaster. The drainage material should be covered by an impermeable fabric with a hole cut through to allow insertion of a PVC flange and pipe for drainage.

4.8 AESTHETICS AND LANDSCAPING

Aesthetics and landscape treatment can be an important serviceability or environmental consideration. This often governs the selection of the overall configuration and appearance of the soil nailed structure. Aesthetics and views on the appearance of the finished soil nailed structure can be subjective and are often a matter of personal preference. In principle, the appearance of a soil nailed structure should be compatible with, and cause minimal visual impact to, the existing environment. The existing vegetation in the vicinity of the soil nailed structure, such as mature trees, should be preserved wherever practical. The as-constructed soil nailed structure should appear as natural as practicable.

The primary objective of landscape treatment of soil nailed structures is to make them both visually acceptable and sustainable in the long term. Designers should consider the following when integrating landscape treatments with engineering works in a soil nailed structure:

- *Minimize physical impact*: addressing the effect of the proposed engineering works on existing vegetation, natural features, and elements of value; achieving best fit to the surrounding landscape and topography; and reducing impact through improved visual treatment of engineering elements and incorporation of planting.
- *Landscape softworks*: planting of vegetation on the face or at the toe, crest, or side edges of a soil nailed structure, in order to create opportunities for ecological habitats to develop, to provide screening, for greening (naturalness), and to improve the environmental condition (e.g. create shade and shelter).

- *Landscape hardworks*: the aesthetic treatment of hard surface covers, and exposed surfaces of the engineering elements and furniture to make their appearance visually acceptable, to create visual interest, and to make the structure compatible with its surroundings.

In most cases, vegetation should be used for surface facing as far as practicable from an aesthetic point of view. If erosion control measures are adopted, one should consider the visual treatment to be adopted for the erosion control technique. On a steep surface profile, the use of an erosion control mat and steel wire mesh may be employed. Shotcrete facing has a rough and irregular appearance which may be improved by trowels or brooms. Furthermore, shotcrete and concrete can be dyed, sculpted, or molded to replicate natural rock formations, stone, or brickwork. A variety of architectural finishes and proprietary greening techniques (see Section 4.9.2) may also be applied to the wall or hard slope facing. Smoothed line top of a nailed wall profile provides a more pleasing appearance than several short abrupt up and down straight line segments. Some special architectural *fascias* comprising precast concrete panels and masonry facing may occasionally be employed to improve the appearance of soil nailed structures.

Where the soil nail heads are exposed, they are visually more acceptable if placed in a regular, rather than random, pattern (see Figure 4.11). The layout of soil nails should be compatible with the preservation and protection of trees on the slope or in the vicinity of the retaining wall. Soil nails and other engineering features should be located away from tree trunks and major roots. Where possible, isolated soil nail heads should be recessed and treated with a matt paint of a suitable color to give a less intrusive visual appearance. An erosion control mat laid over soil nail heads recessed into the slope can also help to reduce their visual prominence. Figure 4.12 shows the appearance of some soil nailed slopes treated by different landscaping techniques.

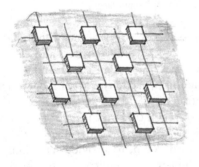

- Adopt regular patterns for soil nail heads to reduce visual impact

- Minimise the size of soil nail heads as far as practicable

Figure 4.11 Regular pattern of soil nail heads.

Figure 4.12 Appearance of soil nailed slopes treated by different landscaping techniques.

4.9 SOIL NAIL HEAD AND FACING

4.9.1 Soil nail head

In most cases, the bond stress at the interface between a soil nail and the ground within the active zone alone is not sufficient to mobilize the full tensile capacity of the soil nail. Some structural contribution from a soil nail head is therefore necessary. The primary function of a soil nail head is to provide a reaction for the individual soil nails to mobilize their tensile forces. Apart from the structural contribution, a side benefit of soil nail heads is to enhance local stability of the ground near the surface as well

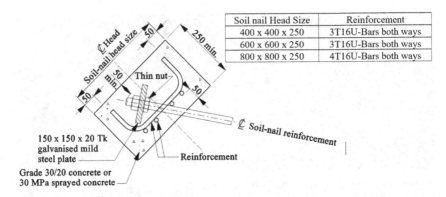

Soil nail Head Size	Reinforcement
400 x 400 x 250	3T16U-Bars both ways
600 x 600 x 250	3T16U-Bars both ways
800 x 800 x 250	4T16U-Bars both ways

Figure 4.13 Typical reinforcement details of a soil nail head in Hong Kong (GEO, 2008). (Note: All dimensions are in millimeters.)

as in between soil nails. Soil nail heads can be recessed into the nailed structure to minimize visual intrusion. Typically, a soil nail head comprises a reinforced concrete pad, steel bearing plate, and hexagonal nut. The bearing plate and nut, which may be embedded within, or bedded onto, the concrete pad, connect the soil nail reinforcement and concrete pad together. A typical detail of a soil nail head being used in Hong Kong is given in Figure 4.13.

Soil nail heads should be designed to provide an adequate safety margin against bearing capacity failure of the soil underneath the nail heads, and against structural failure of the nail heads. One may use Terzaghi's bearing capacity equation (Terzaghi & Peck, 1967) to estimate the size of a square soil nail head. In Hong Kong, the size of soil nail heads has been standardized based on the results of numerical analyses (Shiu & Chang, 2005), which depends on the slope angle, size of nail reinforcement (which reflects the ultimate nail force), and the shear strength parameters of the ground (Table 4.17). Alternatively, the sizes of soil nail heads can be designed using the method recommended by the UK Department of Transport (DOT, 1994) as shown in Figure 4.14.

The effectiveness of soil nail heads in mobilizing tensile forces in soil nails may decrease as the slope angle decreases. Designers should give due consideration in ensuring effective interaction between the soil nail heads and the ground for gentle slopes. For example, designers may consider adopting the details as shown in Figure 4.15 for a gentle slope to enhance the effectiveness of the head.

4.9.2 Facing

A facing primarily serves to provide the soil nailed structure with surface protection, and to minimize erosion and other adverse effects arising

Table 4.17 Recommended sizes of isolated soil nail heads in Hong Kong (GEO, 2008)

Soil Shear Strength Parameters near the Slope Surface		45° ≤ Slope Angle < 55° Diameter of Soil nail Reinforcement (mm)			55° ≤ Slope Angle < 65° Diameter of Soil nail Reinforcement (mm)			Slope Angle ≥ 65° Diameter of Soil nail Reinforcement (mm)		
ϕ'	c' (kPa)	25	32	40	25	32	40	25	32	40
34°	2	800	800	800	600	600	800	600	600	800
	4	600	800	800	600	600	800	600	600	800
	6	600	800	800	400	600	800	400	600	600
	8	600	600	800	400	600	800	400	600	600
	10	400	600	800	400	600	600	400	600	600
36°	2	600	800	800	600	600	800	600	600	800
	4	600	800	800	400	600	800	400	600	800
	6	600	600	800	400	600	800	400	600	600
	8	400	600	800	400	600	600	400	600	600
	10	400	600	800	400	600	600	400	400	600
38°	2	600	800	800	400	600	800	600	600	600
	4	600	600	800	400	600	800	400	600	600
	6	400	600	800	400	600	600	400	600	600
	8	400	600	800	400	600	600	400	400	600
	10	400	600	800	400	400	600	400	400	600
40°	2	600	600	800	400	600	800	600	600	600
	4	400	600	800	400	600	600	400	600	600
	6	400	600	800	400	600	600	400	400	600
	8	400	600	600	400	600	600	400	400	600
	10	400	600	600	400	400	600	400	400	600

Notes: (1) Dimensions are in millimeters unless stated otherwise.

(2) Only the width of the square soil nail head is shown in the table.

(3) The minimum thickness of the soil nail head should be 250 mm.

Determine the size of square soil nail head (w) :

$$w = \sqrt[3]{\frac{T}{\eta}}$$

$$\eta = \frac{\gamma(1-r_u)\tan\beta_s\,e^{3(\frac{\pi}{4}-\frac{\phi'}{2}+\alpha_s)\tan\phi'}}{2\cos(\frac{\pi}{4}+\frac{\phi'}{2})\,(1-\sin\phi')}$$

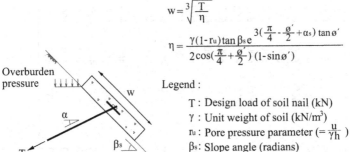

Legend :

T : Design load of soil nail (kN)

γ : Unit weight of soil (kN/m^3)

r_u : Pore pressure parameter ($= \frac{u}{\gamma h}$)

β_s: Slope angle (radians)

ϕ': Angle of shear resistance of soil under effective stress condition (radians)

α : Inclination of soil nail (radians)

u : Pore water pressure (kPa)

h : Depth of overburden directly above the point in question (m)

Figure 4.14 Soil nail head design method recommended by the UK Department of Transport. (Modified DOT, 1994.)

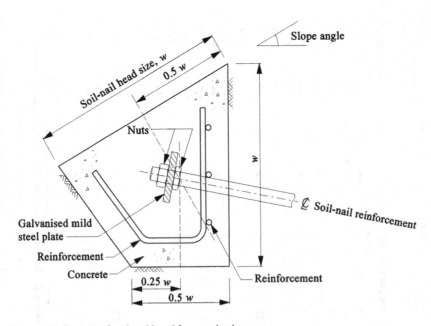

Figure 4.15 Details of soil nail head for gentle slope.

from surface water to the structure. It may be soft, flexible, hard, or a combination of the three. A soft slope facing is normally non-structural, whereas a flexible or hard slope facing can be either structural or non-structural. Flexible structural facings can provide stability to the face of a soil nailed structure by distributing the loads among soil nail heads. These facings allow a certain degree of ground deformation. The function of hard structural facings is similar to that of flexible structural facings but with less ground deformation. Both flexible and hard structural facings provide structural connectivity between soil nails, which promotes integral action of the soil nailed slope and enhances local stability of the slope surface. Table 4.18 summarizes the characteristics of the three types of facings.

Table 4.18 Characteristics of different types of facings

Facings		Characteristics
Soft	Non-structural	These are primarily vegetated covers with synthetic erosion control mat, geogrids, cellular geofabrics, or light steel wire mesh. Although they serve no structural purpose, they can retard surface water from infiltrating into the structure and provide a natural and pleasant environment. Such facings are normally applied to soil nailed structures, of less than about 50°. The long-term performance of such facings depends on the growth and maintenance of the vegetation. As such, planting species should be carefully selected with due regard to local climate and site setting.
Flexible	Non-structural	These facings share the same characteristics as the flexible structural facings except they have no structural contribution to the stability of the soil nailed structures and they allow a higher degree of ground deformation and bulging of the structure.
	Structural	These facings enable transfer of loads from the soil nails to the nail heads. In addition, they allow a certain degree of ground deformation and minor bulging of the soil nailed structures. They are normally composed of heavy-duty coated metallic meshes with prestressed loads. Such facings can also accommodate vegetation to improve their appearance. Proprietary products are available in the market. A limitation of the facings is that the prestressed loads should be monitored and maintained in the long term if they are needed to ensure stability of the soil nailed structure.
Hard	Structural	These facings share the same characteristics as flexible structural facings except that less ground deformation is involved. They provide an effective cover to prevent surface water from infiltrating into the soil nailed structures. Drainage provisions, such as weep holes, should be provided so that excessive groundwater pressure would not be built up behind the facings. Such facings normally comprise sprayed concrete with steel mesh, concrete panels, stone pitching, etc. They are suitable for use in steep soil nailed structures (e.g. > 50°). Landscaping provisions should be provided to improve their appearance.

The type of facing to be adopted depends on structural and aesthetic considerations. A hard facing may enhance the mobilization of tensile capacity of soil nails and prevent local failure of soil between soil nails, whereas a soft or flexible facing can accommodate greening measures, albeit sometimes with somewhat larger ground deformation. One should note, however, that minor failures may occur in between nail heads if a soft cover is adopted for a soil nailed structure, particularly if the slope profile is steep. For example, according to Hong Kong's experience, failures on vegetated soil nailed slopes have occurred occasionally, although they mainly involved local and minor erosion or detachment from shallow depths in the near-surface material within the active zone.

A variety of proprietary greening techniques has been developed which can provide vegetated faces on a steep hard surface cover. Amongst the various greening techniques, surface mulching system of different forms is the most popular technique applied on hard surfacing in Hong Kong. Table 4.19

Table 4.19 Common proprietary surface mulching systems for slope works in Hong Kong (GEO, 2011)

System	Characteristics	Application
Sprayed mulching systems	These systems involve one or more layers of mulch or soil-mix applied onto the slope surface. Mulch and soil-mix vary in type, composition, and additives. They generally have little tensile, compressive, and shear strength. Some systems use a layer of steel wire mesh and/or one or more layers of erosion control mat to strengthen the mulch or soil-mix layer.	Most sprayed mulch systems can be applied on uneven slope profiles to form smooth finished surfaces, hence are suitable for slopes which may have an irregular surface. Cracks in the mulch or soil-mix layer associated with desiccation and down-slope movement of the mulch or soil-mix have been observed on some slopes in the longer term.
Cellular systems	These are systems with compartments or panels to contain the mulch or soil-mix on the slope face. In some of the products, the mulch or soil-mix is embedded into the slope, with the top surface exposed for planting purposes. The panels may be encased in a wire mesh containment system.	Some products allow vegetation to be pre-grown in the mulch or soil-mix panels (in the nursery) before being installed on the slope face. These provide an instant greening effect. However, the small isolated panels of mulch or soil-mix may dry up and result in poor plant growth or plant failure. Without a dense vegetation cover, the panels and wire mesh can become visible and unsightly.
Reinforced soil systems	These are similar to sprayed mulch systems but involve spraying the soil-mix onto the slope face while simultaneously applying continuous synthetic fibers to the soil-mix for strengthening.	The use of reinforcement permits a thicker layer of soil-mix to be constructed on steep slopes. The thicker soil layer can sustain plants longer in dry conditions and can also support the growth of small shrubs.

summarizes the characteristics and application of these surface mulching systems for slope works in Hong Kong.

Designers should provide suitable facings with due regard given to the stability and cost-benefit of the options. For facings requiring long-term maintenance, designers should duly consider the maintenance requirements throughout the design life of the soil nailed structure. Consideration should be given to the following factors in the choice of facing:

- Effectiveness in providing surface protection and erosion control.
- Effectiveness in redistributing soil nail forces between soil nails as deformation of the structure takes place.
- Effectiveness in preventing local failures between soil nail heads.
- Ease of construction.
- Time needed for vegetation establishment.
- Maintenance requirements.
- Initial and maintenance costs.
- Aesthetics.

Figure 4.16 presents the guidelines currently in use in Hong Kong for different types of facings based primarily on the gradient of a soil nailed structure. Due considerations have been given to structural and aesthetic requirements, as well as the cost of long-term maintenance, in formulating these guidelines.

4.10 MATERIALS

4.10.1 Reinforcement

The tensile strength provided by soil nail reinforcement plays an important role in ensuring the stability of a soil nailed structure and hence the suitability and durability of reinforcement should be critically assessed. Materials that are commonly used for soil nail reinforcement include coated or uncoated steel, stainless steel, and fiber composite materials. The reinforcement could be in the form of different cross-sections, such as solid or hollow bars, angle bars, and strips. Figure 4.17 shows some of the typical soil nail reinforcement.

Amongst the various types of materials, steel is the most common material used for soil nail reinforcement because of its high tensile strength and ductile behavior. Steel reinforcing bars are commonly high tensile bars of a nominal diameter of 16 mm to 50 mm and with a tensile strength of 400 MPa to 550 MPa. The common practice of soil nailing in Hong Kong involves the use of solid steel bars with tensile strength of 500 MPa, whereas in the United States, bars of 420 MPa and 520 MPa are common. Under special circumstances, steel bars of higher tensile strength may be

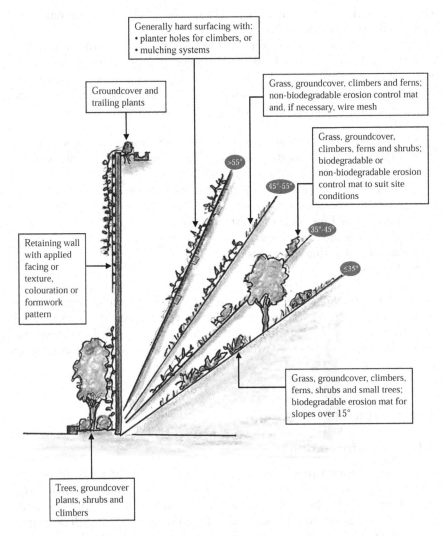

Generally hard surfacing with:
• planter holes for climbers, or
• mulching systems

Groundcover and trailing plants

Grass, groundcover, climbers and ferns; non-biodegradable erosion control mat and, if necessary, wire mesh

Grass, groundcover, climbers, ferns and shrubs; biodegradable or non-biodegradable erosion control mat to suit site conditions

Retaining wall with applied facing or texture, colouration or formwork pattern

>55°

45°-55°

35°-45°

≤35°

Grass, groundcover, climbers, ferns, shrubs and small trees; biodegradable erosion mat for slopes over 15°

Trees, groundcover plants, shrubs and climbers

Figure 4.16 Guidelines for different types of facings for a soil nailed structure in Hong Kong (GEO, 2011).

employed. For example, bars with a tensile strength of 665 MPa and as high as 1,035 MPa are occasionally used in the United States. However, use of such high tensile strength reinforcement should be carefully assessed as the material tends to be more brittle and relatively more susceptible to stress corrosion than steel of lower tensile strength. Apart from normal reinforcing bars with ribs, steel bars with continuous spirally deformed ribbing (continuous threaded bars) are also used in the United States. Soil nail reinforcing bars are generally continuous without connectors. Reinforcement

(a) Galvanized steel bars   (b) Solid/clad stainless steel bars

(c) Epoxy coated steel-strand tendon   (d) FRP bars/strips

Figure 4.17 Various types of soil nail reinforcement.

connectors, also known as couplers, may be used in case extension of the length of soil nails is necessary.

Stainless steel is another feasible option for soil nail reinforcement, which is basically a group of steel alloys that have more than 10.5% chromium. In general, stainless steel is classified into four categories according to their microstructure, namely (i) ferritic, (ii) martensitic, (iii) austenitic, and (iv) duplex. Due to its higher cost and susceptibility to localized corrosion, stainless steel is less commonly used than coated steel for soil nail reinforcing bars. As a result of recent advances in technology, ordinary high strength tensile bars can now be coated with stainless steel (known as clad stainless steel bars) such that the material becomes more cost competitive.

Although steel soil nail reinforcing bars can generally provide adequate tensile strength, under some circumstances twin-bar nails or other proprietary high-strength reinforcement may need to be used in order to provide higher nail capacity. In highly aggressive grounds, additional measures such as corrugated plastic sheathing is required as a corrosion protection measure to the steel reinforcement. Apart from the higher construction cost associated with such measures, other limitations include potential difficulties during installation and longer construction time should be considered. With the advances made in material science, there is an increasing trend of using composite materials comprising synthetic fibers embedded in a polymeric resin, also known as fiber reinforced polymers or fiber reinforced plastics (FRP), as an alternative to steel reinforcement in civil engineering works. The main impetus for considering FRP as an alternative to steel reinforcement is their good corrosion resistance. Fibers commonly used in composites for civil engineering works include carbon FRP (CFRP), glass FRP (GFRP), and aramid FRP (AFRP). Others that are not widely used as yet include polyethylene, quartz, boron, ceramics, and natural fibers. CFRP reinforcement has been applied to slope stabilization works in some countries such as the United States, the UK, Spain, Greece, Japan, and Korea.

FRP reinforcement is anisotropic in nature and is characterized by a high tensile strength in the direction of the reinforcing fibers. However, FRP reinforcement does not exhibit plastic yield behavior before failure as that in steel reinforcement upon loading. Figure 4.18 shows the stress-strain curves of typical high yield steel bar and CFRP bar. One may note that rupture of the CFRP bar occurs at a strain of about 2% without exhibiting any

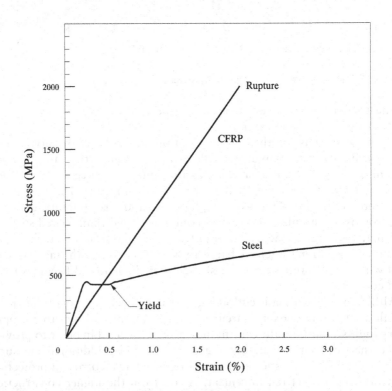

Figure 4.18 Stress-strain curves of typical high yield steel bar and CFRP bar.

plastic yield behavior, whereas the steel reinforcement continues to yield plastically beyond 3% strain. In practice, a larger material safety factor for FRP reinforcement than that for steel, or provision of redundant structural members, could be used to improve the safety margin to compensate for the lack of ductility in FRP material (e.g. IStructE, 1999; ACI, 2003).

In conclusion, the following factors should be considered in selecting a suitable material for reinforcing bars:

- Degree of ductility or brittleness.
- Tensile strength.
- Susceptibility to corrosion.
- Availability in market and time of delivery.
- Cost.

4.10.2 Connection components

Couplers are sometimes used for long soil nails. In principle, these coupling devices should fulfil the same requirements as the soil nail reinforcement in respect of tensile strength, ductility, durability, etc. From the durability perspective, the locations of couplers are susceptible to corrosion and should be considered during the detailed design. Various types of couplers are available as described below:

4.10.2.1 Couplers with threads

This is the most common type of coupler in which threads are rolled or forged on the reinforcement with the threaded portion of a diameter equal to or greater than the main reinforcement. The threaded ends are screwed into the couplers.

4.10.2.2 Couplers with shear bolts

Unlike couplers with threads, this type of coupler does not require the reinforcement to be threaded. It consists of a cylindrical steel coupler with a line of "lockshear" bolts running along its length. The two sections of reinforcement to be joined together are placed in the coupler and the bolts are tightened up, forcing the sections against hardened steel locking strips until the specified torque has been reached.

4.10.2.3 Metal sleeves swaged onto reinforcement

A seamless malleable steel sleeve is slipped over the abutting ends of two reinforcing bars. The sleeve is then swaged and deformed onto the ends of the reinforcing bars using a hydraulic press in which the sleeve is compressed laterally and grips the ribs on the bars.

4.10.3 Centralizers

Centralizers, also known as spacers, are devices that are installed at regular intervals, typically 1.5 m to 2.5 m, along the full length of a soil nail bar to ensure a minimum thickness of grout cover to the bar. Centralizers should be made of non-corrodible materials with no deleterious effects on the soil nail reinforcement, such as polyvinyl chloride (PVC) and other synthetic materials. They should be of sufficient strength and secured properly so that they remain attached to the reinforcement during soil nail installation. This is particularly important for self-drilled soil nails whereby the installation involves partially inserting and withdrawing the reinforcement, which may lead to damage of some of the centralizers.

4.10.4 Cement grout mix

Cement grout, typically made of Portland cement and water, is placed in the pre-drilled hole after insertion of soil nail reinforcement. The cement grout annulus provides intimate contact between a soil nail and the ground. It serves the primary function of transferring stresses between the ground and the soil nail reinforcement. It also provides a nominal level of corrosion protection to the reinforcement. Although cement grout for soil nails is commonly a neat cement grout, sand-cement grout is sometimes used for economic reasons. The characteristics of cement grout can have a significant effect on the ultimate bond strength at the grout-ground interface and the durability of soil nails. The following should be considered in designing the cement grout mix:

- The water-cement ratio should range from 0.4 to 0.5. Occasionally, a thicker grout may be used for the purpose of controlling grout leakage in highly permeable ground or highly fractured rock.
- The chloride content should be kept to a low level (e.g. this is taken to be less than 0.1% of the weight of cement in Nordic countries).
- The grout should be thoroughly mixed and flowability should be adequate to ensure the drillhole is completely filled with the grout and that no water or air is encapsulated in the drillhole.
- The grout should have low tendency for bleeding such that separation and settling of grout are avoided.
- The strength of the cement grout should be high enough to facilitate stress transfer between a soil nail reinforcement and the ground. Typical minimum 28-day unconfined compressive strength required ranges from 20 MPa to 30 MPa. In most cases, admixtures are not needed for the grout mix, but plasticizers may be required to improve the workability of the grout, in particular for projects under high temperature climates, or where project constraints dictate that the grout must be pumped over long distances.

4.10.5 Corrosion protection

Apart from cement grout, which can provide a physical barrier preventing direct contact between soil nail reinforcement and corrosive agents, other corrosion protection measures are needed to ensure the durability of soil nails. Depending on the design life of the soil nailed structure and soil aggressivity, various types of corrosion protection measures can be used to improve durability. Common types of corrosion protection measures include direct coatings (e.g. hot-dip galvanizing or fusion-bonded epoxy) and encapsulating the soil nail reinforcement in a continuous corrugated plastic sheathing of high-density polyethylene (HDPE). Provision of sacrificial steel is also a simple form of corrosion protection measure. Heat-shrinkable sleeves made of polyethylene and anti-corrosion mastic sealant material are commonly used to protect couplers. More details on corrosion protection and durability can be found in Chapter 5.

4.11 CONCLUDING REMARKS

This chapter briefly outlines the design process of a soil nailed structure. The objective of the design is to arrive at the required layout and dimensions of soil nails such that the soil nailed structure can perform its intended functions in ways that are economical, safe, and environmental friendly throughout its design life. Basically, the design process can be divided into preliminary design stage and detailed design stage. A preliminary design is a bridge between design conception and detailed design, during which the design requirements, potential failure modes, loading conditions, site constraints, buildability, etc., should be identified and examined. It is intended to come up with a preliminary layout and dimensions of the soil nails for further assessment and refinement during the detailed design stage when more site-specific information becomes available. Unlike the preliminary design stage, a more comprehensive and detailed assessment should be carried out at the detailed design stage, where design model, design approach, method of analysis, and the safety margin should be clearly identified and defined. The design, either involving empirical approach or analytical analysis, is then carried out to ensure compliance with safety, environment, economic, and durability requirements. It is envisaged that the final layout and dimensions of the soil nails, drainage, aesthetics and landscape treatment, together with the materials and corrosion protection measures to be used, could be determined upon completion of the detailed design.

REFERENCES

ACI (American Concrete Institute) (2003). Guide for the Design and Construction of Concrete Reinforced with FRP Bars (ACI 440.1R-01). American Concrete Institute, Farmington Hills, MI, USA, 41 p.

Bishop, A.W. (1955). The use of the slip circle for the stability analysis of slopes. *Géotechnique*, vol. 5, no. 1, pp. 7–17.

Bock, H. (2006). Common ground in engineering geology, soil mechanics and rock mechanics: Past, present and future. *Bulletin of Engineering Geology and the Environment*, vol. 65, pp. 209–216.

BSI British Standards Institution (2013). Code of Practice for Strengthened/ Reinforced Soils. Part 2: Soil Nail Design (BS 8006: Part 2: 2011). British Standards Institution, London, UK, 104 p.

Clouterre (1991). French National Research Project Clouterre: Recommendations Clouterre (English Translation 1993). Federal Highway Administration, US Department of Transportation, Washington, DC, USA, Report No. FHWA-SA-93-026, 321 p.

DOT (Department of Transport) (1994). Design Methods for the Reinforcement of Highway Slopes by Reinforced Soil and Soil Nailing Techniques. In *Design Manual for Roads and Bridges, Part 4 (HA 68/94)*. The Highways Agency, Department of Transport, UK, 112 p.

GEO (Geotechnical Engineering Office) (2008). *Guide to Soil Nail Design and Construction (Geoguide 7)*. Geotechnical Engineering Office, Civil Engineering and Development Department, HKSAR Government, Hong Kong, 97 p.

GEO (2011). *Technical Guidelines on Landscape Treatment for Slopes (GEO Publication No. 1/2011)*. Geotechnical Engineering Office, Civil Engineering and Development Department, HKSAR Government, Hong Kong, 217 p.

IStructE (The Institution of Structural Engineers) (1999). *Interim Guidance on the Design of Reinforced Concrete Structures Using Fibre Composite Reinforcement*. The Institution of Structural Engineers, UK, 116 p.

Janbu, N. (1954). Applications of composite slip surfaces for stability analysis. In Proceedings of the European Conference on the Stability of Earth Slopes, Stockholm, vol. 3, pp. 39–43.

Krahn, J. (2003). The 2001 R.M. Hardy Lecture: The limits of limit equilibrium analyses. *Canadian Geotechnical Journal*, vol. 40, no. 3, pp. 643–660.

Lazarte, C.A., Elias, V., Espinoza, R. D. & Sabatini, P.J. (2003). Geotechnical Engineering Circular No. 7: Soil Nail Walls. Federal Highway Administration, US Department of Transportation, Washington, D.C., USA, Report No. FHWA0-IF-03-017, 182 p.

Lazarte, C.A., Robinson, H., Gomez, J.E., Baxter, A., Cadden, A. & Berg, R. (2015). Geotechnical Engineering Circular No. 7: Soil Nail Walls: Reference Manual. Federal Highway Administration, US Department of Transportation, Washington, D.C., USA, Report No. FHWA-NHI-14-007, 385 p.

Long, J.H., Sieczkowski Jr., W.F., Chow, E. & Cording, E.J. (1990). Stability analyses for soil nailed walls. In *Design & Performance of Earth Retaining Structures*. Geotechnical Special Publication No. 25 (edited by Lambe & Hansen), ASCE, Reston, VA, pp. 676–691.

Morgenstern, N.R. & Price, V.E. (1965). The analysis of the stability of general slip surfaces. *Géotechnique*, vol. 15, no. 1, pp. 79–93.

Phear, A., Dew, C., Ozsoy, B., Wharmby, N.J., Judge, J. & Barley, A.D. (2005). Soil Nailing: Best Practice Guidance. Construction Industry Research & Information Association, London, UK, CIRIA Report No. C637, 286 p.

Shiu, Y.K. & Chang, G.W.K. (2005). Soil Nail Head Review. Geotechnical Engineering Office, Civil Engineering and Development Department, HKSAR Government, Hong Kong, GEO Report No. 175, 106 p.

Shiu, Y.K. & Chang, G.W.K. (2006). Effect of Inclination, Length Pattern and Bending Stiffness of Soil Nails on Behaviour of Nailed Structures. Geotechnical Engineering Office, Civil Engineering and Development Department, HKSAR Government, Hong Kong, GEO Report No. 197, 116 p.

Shiu, Y.K., Chang, G.W.K. & Cheung, W.M. (2007). Review of Limit Equilibrium Methods for Soil Nail Design. Geotechnical Engineering Office, Civil Engineering and Development Department, HKSAR Government, Hong Kong, GEO Report No. 208, 107 p.

Singh, S. & Shrivastava, A.K. (2017). Effect of soil nailing on stability of slopes. *International Journal for Research in Applied Science & Engineering Technology*, vol 5, no. X, pp. 752–763.

Singh, V.P. & Babu, G.L.S. (2010). 2D Numerical simulations of soil nail walls. *Geotechnical and Geological Engineering*, vol. 28, no. 4, pp. 299–309.

Spencer, E. (1967). A method of analysis of embankments assuming parallel inter-slice forces. *Géotechnique*, vol. 17, no. 1, pp. 11–26.

Terzaghi, K. & Peck, R.B. (1967). *Soil Mechanics in Engineering Practice* (2nd Edition). Wiley, New York, 729 p.

Chapter 5

Durability

5.1 INTRODUCTION

Corrosion, which comes from the Latin word *"corrodere,"* means "to attack" or "to chew away" material as a result of chemical and/or physical interaction between the material and the environment. Soil nails are commonly installed in the ground that is subjected to variations in environmental conditions, such as seasonal rise of groundwater and freeze-thaw cycles, as well as exposed to different physical and chemical soil compositions. As the ground is a complex physical and chemical environment that can vary both spatially and temporally, potential degradation of metallic soil nails through corrosion may affect the long-term performance of a soil nailed structure. Durability is therefore an essential design requirement of a soil nailed structure. Durability of a soil nailed structure depends primarily on the corrosion potential of the ground and the ability of the soil nailed structure to withstand corrosion attack from the surrounding soils. The former is related to the physical and chemical composition of the ground and groundwater as well as the change in environmental conditions over time, whereas the latter depends on the material used for the soil nails, the degree of corrosion protection employed, and the long-term integrity of the protection measures after construction.

Based on historical field observations and research findings, the lifespan of steel soil nails can vary greatly from months to decades, depending on the corrosivity of the ground and the degree of corrosion protection provided to the soil nails. Past experience also suggests that although the ground conditions which exist at the time of soil nail construction could be determined with some accuracy, the potential changes in environmental conditions during the lifespan of a soil nailed structure could be highly uncertain. In view of this uncertainty, it may be prudent in some cases for the design to err on the conservative side in terms of provision of corrosion protection measures.

In this chapter, the mechanism of corrosion, the corrosivity classification of the ground, and the range of corrosion protection measures to soil nails are discussed. To date, as steel is the most commonly used material in soil

137

nailing, the majority of the coverage of this chapter is therefore devoted to steel soil nails. However, with the rapid development of new technology and innovation in material science, some advanced materials with good performance in terms of durability, such as fiber reinforced polymers, are becoming more popular for use in soil nailing. This new area is also covered in this chapter.

5.2 MECHANISM OF CORROSION

Metals or alloys are thermodynamically unstable in the environment and they tend to revert from a temporary high energy state to a natural lower energy state, such as oxide and hydroxide, by reacting with oxygen and water. The process is called corrosion, which is primarily an irreversible deterioration process by chemical or electrochemical means. For corrosion to occur, there must be a difference in electrical potential, oxygen concentration, pH, or electric resistivity between two points that are electrically connected in the presence of a non-metallic conductor existing in solution (electrolyte). The reaction involves movement of ions from the surface of a metal or alloy (anode) to the surface of another metal (cathode) through the electrolyte. In the case of steel soil nails in the ground, the electrolyte is the soil pore water which contains both oxygen and often dissolved salts. Invariably, corrosion will result in a detrimental effect to the long-term performance of soil nailed structures.

Corrosion takes place at the reactive anode areas and the non-reactive cathode areas on the steel surface. The processes and the chemical equations involved are illustrated in Figure 5.1.

Anodic and cathodic reactions are generally referred to as "half-cell reactions." The anodic reaction is the oxidation process which results in loss of metal, whereas cathodic reaction is a reduction process which leads to reduction of dissolved oxygen forming hydroxyl ions. In other words, oxygen and water have to be present simultaneously in order for metals or alloys to corrode. Corrosion can also occur due to attack by deleterious substances present in the soils. For example, steel is corroded rapidly by hydrochloric acid:

$$Fe + 2HCl \rightarrow FeCl_2 + H_2$$

There are many types of corrosion, viz. general or uniform, localized or pitting, stray-current, bacterial, hydrogen embrittlement, crevice and galvanic corrosion. In the context of soil nailing, uniform, pitting, and stray-current corrosions are the most common types of corrosion with respect to reinforcement. Table 5.1 summarizes the characteristics of these major types of corrosion in soil nails. Uniform corrosion is characterized by corrosive attack proceeding evenly over the entire surface area or a large fraction of

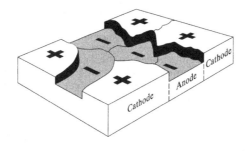

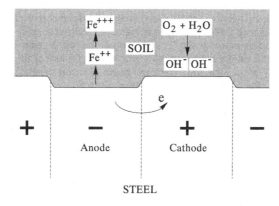

At Anode	$Fe \rightarrow Fe^{++} + 2e^-$
	$Fe^{++} \rightarrow Fe^{+++} + e^-$
At Cathode	$O_2 + 2H_2O + 4e^- \rightarrow 4OH^-$
Combined reaction	$4Fe + 3O_2 + 2H_2O \rightarrow 2Fe_2O_3 \cdot H_2O$ (Rust)

Figure 5.1 Schematic diagram showing electrochemical reactions involved in steel corrosion. (Modified from British Steel Corporation.)

the area of reinforcement. In general, metal loss due to uniform corrosion is not significant as long as the detrimental conditions promoting corrosion are absent. Pitting corrosion, on the other hand, is caused by depassivation of a small area, involving a localized form of corrosion that leads to the creation of small holes in the reinforcement. In the case of soil nailing, differential aeration cells may occur when a steel reinforcement passes through soils of very different permeabilities (e.g. from gravel to silt or clay). The portion of soil nail surrounded by high concentration of dissolved oxygen would form a cathodic area, while the other part with low concentration of dissolved oxygen would form an anodic area. Pitting usually occurs at locations close to the boundary between the two areas. Stray-current corrosion is caused by the soils which are exposed to stray current for a considerable period of time and hence exacerbates the flow of ions between metals.

Table 5.1 Characteristics of major types of corrosion in soil nails

Types of Corrosion	Characteristics
General (Uniform) Corrosion	This type of corrosion occurs on the surface of steel reinforcement resulting in a uniform thin layer of rust. However, once this thin passive rusting film is formed on the reinforcement, the oxidation rate will decrease. This rusting layer effectively inhibits further corrosion of the reinforcement. If the conditions that inhibit corrosion are absent, general corrosion will likely be significant over a typical period of several decades.
Localized Corrosion (Pitting)	Localized corrosion or pitting develops in small pits or crevices, which promote non-homogeneous concentrations of ions, an increase in electrochemical potential, and further localized corrosion. Once a pit is created, if sufficient oxygen is available, localized corrosion tends to propagate deeper, the cross-section of the reinforcement is reduced at that location, and overstressing or eventual failure of the reinforcement may occur. Thus, the extent and depth of pitting depend on the availability of oxygen to continue the reaction.
Stray-Current Corrosion	Stray currents are electric currents flowing from external sources. When there is prolonged exposure to stray currents, they tend to exacerbate the flow of ions between metallic parts and hence accelerate the corrosion process. Stray currents tend to cause corrosion where the ion flow is strongest at locations where the current leaves the structure and enters the ground or water electrolyte. Besides the distance to the stray current source, other factors may affect the rate of corrosion of metals subject to stray currents, including source voltage, characteristics of the metallic surface (e.g. steel, zinc), type of electrolyte (e.g. water, water contaminated with other fluids), presence and concentration of substances in the electrolyte (e.g. salts such as carbonates and chlorides), nature of anode/electrolyte interface, and other environmental factors.

Generally speaking, pitting corrosion is more hazardous than uniform corrosion in that it is more difficult to predict and design for.

5.3 CORROSION RATE

5.3.1 Factors promoting corrosion

There are many factors that would promote corrosion of steel reinforcement and the associated components in soil nails. These include, *inter alia*, the availability of oxygen and moisture, temperature, carbonation, and entry of acidic pollutants, chloride ions that can reach the reinforcement either from cement grout ingredients or from the external environment, presence of stray currents, composition of soil, construction practice, and permeability of or presence of cracks in cement grout, if any, encasing the reinforcement.

Soil aggressivity or corrosivity is considered the most appropriate and important indicator in reflecting collectively the corrosion potential of a soil. Soil aggressivity is not a measurable parameter. Therefore, in the evaluation of soil aggressivity, a number of critical factors characteristic of the soil are usually employed. The aggressivity of soils can vary over a wide range because of the variability of environmental conditions as well as soil compositions and properties. In general, the corrosion rate of steel soil nails in a soil depends primarily on the soil's physical and chemical characteristics. The relevant physical characteristics are those that control the permeability of the soil to air and water. These include grain size, permeability, and moisture content of the soil. Fine-grained soils (silts and clays) are potentially more corrosive than coarse-grained soils (sands and gravels) in which there is greater circulation of air and less water-retention capacity. The relevant chemical characteristics are those that determine the ability of the soil to act as an electrolyte for the development of local corrosion cells. These include alkalinity, acidity, concentrations of oxygen and dissolved salts, and organic matter and bacteria content. These factors affect electrical resistance, which is accepted as a key parameter for reflecting the corrosivity of a soil (King, 1977; Eyre & Lewis, 1987; Fontana, 1987). Stray currents, where present, can also influence the corrosion rate. These may initiate corrosion or even accelerate existing corrosion processes.

In summary, soil aggressivity is a useful indicator that reflects collectively the corrosion potential of a soil. It can be determined by reference to a set of measurable parameters, including soil type, water content, pH value, redox potentials, microbiological activity, anion and cation levels, electrical resistivity, etc. The key parameters that would affect the aggressivity of a soil are summarized in Table 5.2.

5.3.2 Evaluation of corrosion rate

Broadly speaking, corrosion is a time-dependent process which can be divided into two distinct stages, namely the (i) initiation and (ii) propagation stages. The initiation stage is the period of time from soil nail installation to the moment when the reinforcement is no longer passivated, i.e. the start of corrosion or depassivation (passivation refers to the conditions in which the material is less affected by corrosion, see Section 5.5.2 on passivation of steel surface). The propagation stage is the period of time during which corrosion of the depassivated reinforcement takes place until non-performance of the soil nail. The process is depicted schematically in Figure 5.2.

A comprehensive source of information on underground corrosion, comprising extensive field exhumation and testing on metal pipes and sheet steel, is a range of study programs launched by the US National Bureau of Standards (NBS) as early as 1910 (Romanoff, 1957). The results of the studies are widely reported in literature relating to corrosion of metals in soil

Table 5.2 Major factors affecting soil aggressivity

Corrosion Factors	Details
Physical Properties	The physical properties of soils that are of importance to soil corrosion are those which determine the permeability of the soil to air and water. Soils with a coarse texture, such as sands and gravels, permit free circulation of air and water. The corrosion under such conditions is broadly similar to that under exposure to the atmosphere. On the other hand, clayey and silty soils are generally characterized by a fine texture and high water-retention capacity, resulting in poor aeration and poor drainage. These characteristics tend to reduce general corrosion but make local pitting of the steel more likely to occur and can lead to an increase in corrosion rate by keeping the surface wet for a longer period. The pertinent physical soil properties include particle size distribution, Atterberg limits, moisture content, etc. According to King (1977), at 20% moisture content and below, the rate of diffusion of gaseous oxygen could be very high and could result in high corrosion rates. Elias et al. (2009) suggested that when the moisture content of a soil is greater than 25% to 40%, the rate of general corrosion is increased. Below this value, pitting corrosion is more likely.
Soil Resistivity	Soil resistivity is defined as the inverse of electrical conductivity of the soil. This is a very important parameter that governs the corrosivity of a soil. It is a measure of the ability of a soil to act as an electrolyte and is related to the development of local corrosion cells as well as ease of transmission of stray currents. In general, corrosion rate increases as soil resistivity decreases. However, if the resistivity is high, localized corrosion (pitting) rather than general corrosion is more likely to occur. The resistivity of a soil depends on a number of factors, in particular moisture content, salt content, degree of compaction, temperature, etc.
Stray Current	Stray currents are present in the ground as a result of electrical leaks, presence of transit rail systems, or failure to provide positive and permanent electrical earthing. Stray currents can be an additional trigger of corrosion in a soil nailing system. In general, the effects of stray currents decrease rapidly with distance from the source and become negligible at a distance of about 30 to 60 m.
Redox Potential	The oxidation and reduction potential of a soil depends on the relative proportion of oxidizing and reducing agents in the soil. Redox potential reflects the tendency of a soil to support microbiological activity. The more reducing agents there are in the soil, the lower the level of oxygen and potentially the activity of sulfate reducers will be greater. Redox potentials are measured in the field by measuring the potential of a platinum electrode using a calomel reference electrode.
pH Value	The soil pH value represents the hydrogen ion (H^+) concentration in solution in the soil water. Soil resistivity as mentioned above indicates the ability of a soil to act as an electrolyte for electrochemical corrosion. However, corrosion can also be caused or enhanced by chemical attack and the pH value measurement is used to assess such attack potential. Soils that are strongly acidic (pH < 4) or strongly alkaline (pH > 10) are generally associated with significant corrosion rates.

(Continued)

Table 5.2 (Continued) Major factors affecting soil aggressivity

Corrosion Factors	Details
Soluble Salts	The amount of dissolved inorganic solutes in the soil water is directly proportional to the solution's electrolytic conductivity. This electrolytic conductivity is the sum of the individual equivalent ionic conductivities times their concentration. The more soluble salts are active participants in the corrosion process. Chlorides, sulfates, and sulfides are identified as major salts in promoting corrosion. However, carbonate acts as a corrosion retarder, which forms an adherent scale on the metal surface and reduces the corrosion rate.
Organic Matter	Micro-organisms also affect the chemical properties of a soil through oxidation and reduction reactions. Bacterial activities tend to reduce the oxygen content and replace oxygen with carbon dioxide. Thus, the microbial growth will convert organic matters in soils to organic acids which, when in contact with metal, will facilitate pitting corrosion.

(e.g. King, 1977; McKittrick, 1978; Fontana, 1987; Elias & Juran, 1991; Zhang, 1996; Elias et al., 2009). In 1922, the NBS started a 10-year program in which galvanized pipe specimens were buried in 47 types of soils with resistivity ranging from 60 ohm-cm to 45,100 ohm-cm and pH values from 3.1 to 9.5. For each type of soil, three to six specimens were buried. These specimens were exposed progressively from 2 years to 17 years after burial. The corresponding average corrosion rates of the zinc coating were found ranging from 0.5 to greater than 40.6 µm/year. Among the 47 average corrosion rates in various soils, most of them (38 out of 47) were found to be below 10 µm per year. The test results also indicated that in most soils, zinc coatings of 85 µm or less were completely destroyed in 10 years' burial, whereas for 130 µm zinc coatings, some of the coating remained on the steel

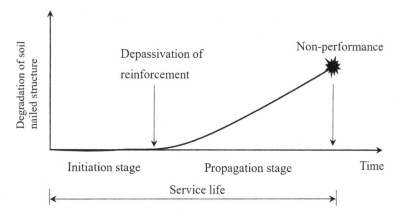

Figure 5.2 Schematic diagram showing the service life of a soil nailed structure.

for at least half of the specimens. Table 5.3 summarizes the results obtained in the NBS investigation.

In 1937, the NBS conducted another test program, using 38 mm steel pipe specimens which were either bare or galvanized with a nominal 130 μm zinc coating. The specimens were buried in 15 soils of different characteristics. The specimens were subsequently exposed at time periods ranging from 4 to 11.2 years after burial. The results in terms of weight loss and maximum pit depths for the test specimens are summarized in Table 5.4. The maximum pitting rates for galvanized steel and plain steel are found to be 1 to 5 times and 1 to 13 times of the corresponding surface average corrosion rates, respectively. The test program concluded that the life of galvanized steel buried in soil would be greatly dependent on the nature of the soil. A nominal 85 μm coating would provide protection for at least 10 years in inorganic oxidizing soils. A 130 μm coating appeared to be adequate (for 10–13 years) in most inorganic reducing soils but would not afford sufficient protection in highly reducing organic or inorganic soils.

Another important finding of the NBS studies is that the rate of corrosion of both steel and zinc decreases with time. There is a rather rapid loss in the first 2 years for both bare and galvanized steels followed by a progressive decrease in the rate of corrosion. Similar observations were made by Darbin et al. (1988) from another independent tests conducted in France. From the NBS test data, the average loss of thickness for steel as a function of time can be expressed empirically by the following equation (Romanoff, 1957).

$$X = Kt^n \tag{5.1}$$

where t is time in years, X is the depth of general corrosion or pit depth in μm at time t, and K and n are constants that depend on the soil and site characteristics (n is always less than 1.0).

For low carbon steels in various soil burial conditions, the NBS established an "n" value varying from 0.5 to 0.6 and a "K" value varying between 150 and 180 μm at the end of the first year (Elias et al., 2009). For galvanized steels, the "K" value ranges between 5 and 70 μm, but the "n" value was not evaluated.

The results of container tests and electrochemical tests reported by Darbin et al. (1988) indicate that for the range of soil fill utilized in reinforced soil structures in France, the "n" value may be taken to be 0.6 for galvanized steel, while the zinc coating is still present, and from 0.65 to 1 for carbon steel once significant corrosion occurs. The "K" value calculated at the end of the first year for galvanized steel was found to vary between 3 and 50 μm, with the higher values consistent with more aggressive soils characterized by lower resistivities and higher concentrations of chlorides and sulfates. Figures 5.3 and 5.4 show the log-log plots of the test data of metal loss versus time for these tests.

Table 5.3 Test results of the National Bureau of Standards Underground Corrosion
Program in 1924 (Romanoff, 1957; Zhang, 1996)

Specimen (See Notes 1 & 2)	Soil Type	Resistivity (ohm-cm)	pH	Average Corrosion Rate of Zinc Coating (μm/year)
1	Allis silt loam, Cleveland, Ohio	1,215	7.0	11.8
2	Bell clay, Dallas Texas	684	7.3	1.5
3	Cecil clay loam, Atlanta, Georgia	30,000	5.2	1.7
4	Chester loam, Jenkintown, Pennsylvania	6,670	5.6	7.9
5	Dublin clay adobe, Oakland, California	1,345	7.0	7.7
6	Everett gravelly sandy loam, Seattle, Washington	45,100	5.9	0.5
7	Maddox silt loam, Cincinnati, Ohio	2,120	4.4	10.8
8	Fargo clay loam, Fargo, North Dakota	350	7.6	3.2
9	Genesee silt loam, Sidney, Ohio	2,820	6.8	5.0
10	Gloucester sandy loam, Middleboro, Massachusetts	7,460	6.6	5.2
11	Hagerstown loam, Loch Raven, Maryland	11,000	5.3	3.7
12	Hanford fine sandy loam, Los Angeles, California	3,190	7.1	2.2
13	Hanford very fine sandy loam, Bakersfield, California	290	9.5	3.7
14	Hempstead silt loam, St. Paul, Minnesota	3,520	6.2	1.1
15	Houston black clay, San Antonio, Texas	489	7.5	1.5
16	Kalmia fine sandy loam, Mobile, Alabama	8,290	4.4	4.2
17	Keyport loam, Alexandria, Virginia	5,980	4.5	14.8 (see Note 3)
18	Knox silt loam, Omaha, Nebraska	1,410	7.3	(see Note 4)
19	Lindley silt loam, Des Moines, Iowa	1,970	4.6	2.9
20	Mahoning silt loam, Cleveland, Ohio	2,870	7.5	4.9
21	Marshall silt loam, Kansas City, Missouri	2,370	6.2	(see Note 4)

(Continued)

Table 5.3 (Continued) Test results of the National Bureau of Standards Underground Corrosion Program in 1924 (Romanoff, 1957; Zhang, 1996)

Specimen (See Notes 1 & 2)	Soil Type	Resistivity (ohm-cm)	pH	Average Corrosion Rate of Zinc Coating (µm/year)
22	Memphis silt loam, Memphis, Tennessee	5,150	4.9	5.2
23	Merced silt loam, Buttonwillow, California	278	9.4	40.6 (see Note 3)
24	Merrimac gravelly sandy loam, Norwood, Massachusetts	11,400	4.5	1.1
25	Miami clay loam, Milwaukee, Wisconsin	1,780	7.2	1.45
26	Miami silt loam, Springfield, Ohio	2,980	7.3	2.9
27	Miller clay, Bunkie, Louisiana	570	6.6	3.9
28	Montezuma clay adobe, San Diego, California	408	6.8	8.8
29	Muck, New Orleans, Louisiana	1,270	4.2	25.5 (see Note 3)
30	Muscatine silt loam, Davenport, Iowa	1,300	7.0	1.9
31	Norfolk fine sand, Jacksonville, Florida	20,500	4.7	0.67
32	Ontario loam, Rochester, New York	5,700	7.3	2.4
33	Peat, Milwaukee, Wisconsin	800	6.8	7.4
34	Penn silt loam, Norristown Pennsylvania	4,900	6.7	(see Note 4)
35	Ramona loam, Los Angeles, California	2,060	7.3	1.3
36	Ruston sandy loam, Meridian, Mississippi	11,200	4.5	1.0
37	St. John's fine sand, Jacksonville, Florida	11,200	3.8	8.7
38	Sassafras gravelly sandy loam, Camden, New Jersey	38,600	4.5	0.85
39	Sassafras silt loam, Wilmington, Delaware	7,440	5.6	(see Note 4)
40	Sharkey clay, New Orleans, Louisiana	970	6.0	4.0
41	Summit silt loam, Kansas City, Missouri	1320	5.5	2.2
42	Susquehanna clay, Meridian, Mississippi	13,700	4.7	3.0

(Continued)

Table 5.3 (Continued) Test results of the National Bureau of Standards Underground
Corrosion Program in 1924 (Romanoff, 1957; Zhang, 1996)

Specimen (See Notes 1 & 2)	Soil Type	Resistivity (ohm-cm)	pH	Average Corrosion Rate of Zinc Coating (μm/year)
43	Tidal marsh, Elizabeth, New Jersey	60	3.1	5.5
44	Wabash silt loam, Omaha, Nebraska	1,000	5.8	1.9
45	Unidentified alkali soil, Casper, Wyoming	263	7.4	7.5
46	Unidentified sandy loam, Denver, Colorado	1,500	7.0	0.7
47	Unidentified silt loam, Salt Lake City, Utah	1,770	7.6	4.3

Notes: (1) Specimen No. refers to the original identification.

(2) Average zinc coating thickness of the 47 specimens is 121 μm.

(3) Zinc coating corroded completely; data included the corrosion of steel.

(4) The average corrosion rate of zinc coating is not available from the original publication.

Using the NBS model, Elias et al. (2009) suggested the following equations for determining the corrosion loss of galvanized steel:

$$X = 25\ t^{0.65}\ (\text{Average}) \tag{5.2}$$

$$X = 50\ t^{0.65}\ (\text{Maximum}) \tag{5.3}$$

and the following equations for corrosion loss of carbon steel:

$$X = 40\ t^{0.80}\ (\text{Average}) \tag{5.4}$$

$$X = 80\ t^{0.80}\ (\text{Maximum}) \tag{5.5}$$

where t is time in years, X is the depth of corrosion in μm.

For reinforced fill structures with selected backfills that meet stringent electrochemical requirements, Elias et al. (2009) proposed that the maximum loss per side due to corrosion may be computed by assuming the following loss rates:

Zinc corrosion rate for the first 2 years	15 μm/yr
Zinc corrosion to depletion	4 μm/yr
Carbon steel loss rate	12 μm/yr

Table 5.4 Test results of the National Bureau of Standards Underground Corrosion Program in 1937 (Romanoff, 1957; Zhang, 1996)

Specimen	Soil Type	Resistivity (ohm-cm)	pH	Galvanized Steel		Steel	
				Average Corrosion Rate (μm/year)	Maximum Pitting Rate (μm/year)	Average Corrosion Rate (μm/year)	Maximum Pitting Rate (μm/year)
1	Acadia clay	190	6.2	22.9	22.5	83	361
2	Cecil clay loam	17,790	4.8	3.8	16.9	16.2	209
3	Hagerstown loam	5,210	5.8	3.9	16.9	19.6	260
4	Lake Charles clay	406	7.1	26.3	36.7	132	409
5	Muck	712	4.8	43.0	180.6	82.7	277
6	Carlisle muck	1,660	5.6	14.1	22.2	35.8*	56.4
7	Rifle peat	218	2.6	77.4*	234	85.5	164
8	Sharkey clay	943	6.8	7.2*	34	20.1	135
9	Susquehanna clay	6,920	4.5	4.3	16.9	25.3	192
10	Tidal marsh	84	6.9	9.6	22.6	51.1	226
11	Docas clay	62	7.5	7.6	28.2	43*	226
12	Chino silt loam	148	8.0	7.6	16.9	33.4	183
13	Mohave fine gravely loam	232	8.0	11.9	16.9	58.8*	409
14	Cinders	455	7.6	58.0	286	151	409
15	Merced silt loam	278	9.4	9.9**	18.5	64	344

Notes: (1) * denotes data averaged over 4 years.

(2) ** denotes data averaged over 11.2 years.

(3) All other data averaged over 9 years.

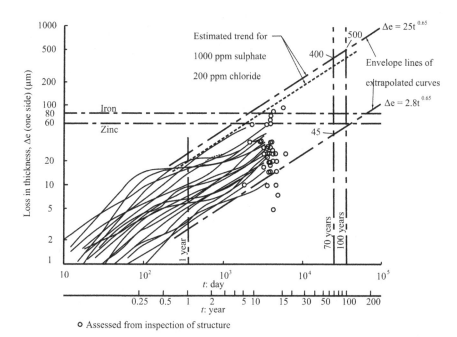

Figure 5.3 Electrochemical tests for quarter-saturated soils which comply with specifications for dry structures (Darbin et al., 1988).

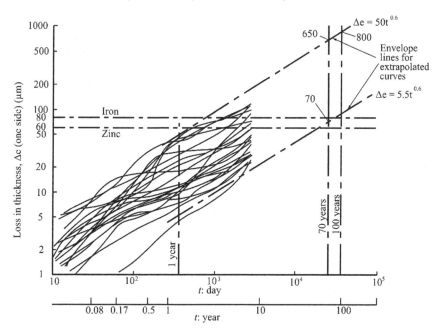

Figure 5.4 Electrochemical tests for saturated and half-saturated soils in accordance with specifications for submerged structures (Darbin et al., 1988).

Table 5.5 Rate of corrosion of galvanizing coating and steel (based on Department of Transport, 1988)

Soil Condition	Rate of Uniform Corrosion of Galvanizing (µm/year)	Depth of Uniform Corrosion of Steel (µm)
Non-Aggressive	4	$22.5\,t^{0.67}$
Aggressive	14	$40\,t^{0.80}$

Note: t is time in years

In the UK, the corrosion allowances for buried corrugated steel structures are specified by the Department of Transport (1988). The assumed corrosion rates for buried structures are given in Table 5.5. They are smaller than the rates suggested by Elias et al. (2009). It is worth noting that the assumed rates were established for uniform corrosion conditions with no allowance for pitting, although corrosion does not normally occur in a uniform manner. Loss of cross-sectional area will be greater where significant pitting or greater localized corrosion occurs as compared to that computed by distributing corrosion losses uniformly over an element (Elias et al., 2009). Surveys by Brady & McMahon (1993) on 46 corrugated steel structures buried in the ground for periods between 16 and 34 years show that corrosion tended to be localized. The NBS data also suggest that pitting depths could be significantly greater than depths due to uniform loss. According to King (1977), the test data from the UK could infer a maximum pit depth of steel of 5.8 mm in 20 years.

Another corrosion related study was carried out by the Swedish Corrosion Institute (Camitz & Vinka, 1989), which included field tests on carbon steel and steel coated with zinc and aluminum-zinc alloy. Specimens were placed in different types of soils in Sweden, above and below the groundwater table, for up to 4 years. The results suggest that the corrosion rate for both carbon steel and zinc-coated steel is in general relatively higher in soils of low pH values. Also, the corrosion of carbon steel and zinc coatings is lower in sands and higher in clays. The corrosion rate of carbon steel specimens above the groundwater table is higher than that below the groundwater table, whereas for zinc coatings the position of the groundwater table has no distinct effects on the corrosion rate. In addition, the corrosion rate was lower for specimens embedded in a homogeneous sand fill than specimens buried directly in *in situ* soils.

It is evident from the results of various studies discussed above that the corrosion rate of buried galvanized steel can vary greatly among different types of soil. According to ZALAS (1989), the performance of galvanized steel elements is best in alkaline and oxidizing soils where a 600 g/m² zinc coating will, in general, give additional life of about 10 years to pipes. Highly reducing soil is the most aggressive and may consume a zinc coating at a rate of more than 13 µm per year. Unprotected galvanized coatings should not be used in environments with a pH of less than 6 or greater than 12.5 (ZALAS, 1989). Within the pH range of 6 to 12.5, the corrosion rate

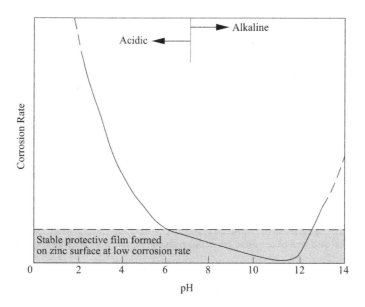

Figure 5.5 Effects of pH on corrosion rate of galvanized coating.

of zinc is relatively low since a stable protective film is formed on the zinc surface. Figure 5.5 shows schematically the relationship between corrosion rate and acidity of soil.

In Hong Kong, the Geotechnical Engineering Office (GEO) initiated a study in 1999 of the long-term durability of soil nails, with a view to developing an improved approach for designing and specifying corrosion protection measures for steel soil nails (Shiu & Cheung, 2003). The study comprised a review of relevant literature on the corrosion of steel in soils and technical guidance documents from various places, including France, the UK, the United States, and Japan, with a view to assessing the effectiveness and reliability of the various corrosion protection measures. As part of the study, six different types of corrosion protection measures that are commonly used for steel soil nails, viz. cement grout, sacrificial steel, stainless steel, sacrificial metallic (zinc) coating, non-metallic coating, and corrugated plastic sheath, were identified for detailed examination. In order to collect long-term data on the corrosion rates of soil nails with these common types of corrosion protection measures, the GEO commenced a program in 2004 to monitor the corrosion conditions of sacrificial soil nails installed in various places with different types of protection provisions for about 50 years. Plain steel bars without any corrosion protection measures were also installed as control specimens. For each type of corrosion protection measure, 12 sacrificial soil nails (six with cement grout annulus and six without), in the form of 1.5 m long, 32 mm diameter high-yield deformed steel bars, were installed

Table 5.6 Investigation schedule of soil nails with different types of corrosion protection measures

| Corrosion Protection | Grout Annulus | No. of Soil Nails | | | | | | |
| | | Installation | Exhumation | | | | | |
		2004	2011	2013	2019	2029	2039	2049
Plain Steel	Yes	6	1	1	1	1	1	1
	No	6	1	1	1	1	1	1
Hot-Dip Galvanizing	Yes	6	1	1	1	1	1	1
	No	6	1	1	1	1	1	1
Epoxy Coated	Yes	6	1	1	1	1	1	1
	No	6	1	1	1	1	1	1
Stainless Steel	Yes	6	1	1	1	1	1	1
	No	6	1	1	1	1	1	1

Note: (1) For each type of corrosion protection measure, 12 nails were installed at each of the four sites.

(2) Diameter of cement grout annulus is 100 mm.

(3) Thickness of galvanizing is 85 μm.

(4) The aggressivity of the sites varies from mildly aggressive to aggressive.

at each of the four different sites in 2004. The corrosion potential of the four sites ranges from mildly aggressive to aggressive. Upon reaching a designated time span, the sacrificial soil nails would be exhumed progressively from 2011 to 2049 as scheduled in Table 5.6 for examination.

The investigation essentially involves re-assessment of soil aggressivity at the time of nail exhumation (i.e. by soil classification tests, chemical tests, resistivity, and redox potential tests), visual examination, and laboratory testing of the exhumed nails. Laboratory testing on the exhumed nails includes carbonation test of cement grout, measurement of zinc coating, weight loss and depth of corrosion, and tensile test of the steel reinforcement bars. So far, two batches of nails have been exhumed and investigated since installation. The aggressivity assessment indicated that the four sites remained the same in terms of corrosivity as that in 2004, i.e. ranging from mildly aggressive to aggressive. For those nails with cement grout annulus, visual inspection indicated that the majority of the grout was intact and in fair condition. They generally showed typical and minor defects, such as minor cracks and voids in the grout annulus. However, the grout annulus of some of the exhumed nails was of poor quality and showed extensive and interconnecting voids. This could possibly be due to poor workmanship of grouting at the time of nail installation.

For those nails embedded in grout of good quality, no sign of corrosion was observed in aggressive ground for up to 7 years. All types of corrosion protection coatings were still found in good condition. This suggests that good quality cement grout contributes as an effective corrosion protection measure. For the plain steel nails, which were protected by hot-dip galvanizing or epoxy coating but without grout annulus, and were in direct contact with aggressive ground, no significant signs of corrosion were observed. The various types of corrosion protection measures (i.e. hot-dip galvanizing and epoxy coating) were found to be intact. In contrast, both uniform and pitting corrosion were observed on the plain steel bars which were in direct contact with aggressive ground. The investigation suggested that the pitting rate of unprotected plain steel in the aggressive ground was about 0.25 mm/year, which is comparable to an average pitting rate of 0.29 mm/year suggested by King (1977). In other words, the pitting rate is about one order greater than the uniform corrosion rate. This observed corrosion rate is similar to that assumed in the design of soil nailed structures in Hong Kong. Furthermore, tensile tests indicated that there was no noticeable reduction of tensile capacity of all exhumed nails.

5.4 SOIL CORROSIVITY CLASSIFICATION

The durability of a soil nailed structure is highly dependent on the aggressivity of the ground, which is a complex physical and chemical environment that can vary considerably both spatially and temporally. As discussed in Section 5.3, soil aggressivity is not a measurable parameter and hence a range of measurable critical factors characterizing the physical and chemical conditions of the soil is necessary in order to assess its aggressivity. It is essential that the assessment of ground aggressivity, preferably in a quantitative manner, be carried out during the site investigation stage. One should, however, bear in mind that the assessment is only relevant to the conditions that exist at the time of construction. The environment conditions that a soil nailed structure is subjected to during its service life could vary considerably and may not be easy to establish with confidence.

In general, adverse environmental conditions that can greatly affect the durability of a soil nailed structure include zones of fluctuating groundwater, high salinity, high or low pH value, high chloride or sulfate content, contaminated soils, and high clay content. In addition, sites with disturbed ground conditions or buried services may pose a risk to the durability of a soil nailed structure. It is generally believed that disturbed ground conditions promote chemical reactions, particularly those resulting from the transport of oxygen, which is more likely to occur than in undisturbed ground (i.e. natural ground). Also, effluents from leakage of buried water-carrying services like sewers can make the ground become aggressive. For example, Hui et al. (2007) reported some cases in Hong Kong where the slopes were affected or contaminated by leakage of buried water-carrying services in the vicinity. In practice, one of the key problems is associated with ageing pipes, and inadequate investigation of the adjacent water-carrying services during the design stage of slope works.

Different assessment schemes have been developed and are being used in different parts of the world for classifying the aggressivity of soils. Each of these schemes emphasizes a slightly different combination of factors as discussed in Section 5.3 and critical values are assigned to the corresponding factors. The soils are generally classified as "aggressive" or "non-aggressive" according to whether the values computed are above or below the critical values.

In France, Clouterre (1991) classified soils into four categories, namely "highly corrosive," "corrosive," "average corrosiveness," and "slightly corrosive." The corrosivity assessment involves the determination of an "overall corrosiveness index," which is based on weightings assigned to four factors, viz. type of soil, resistivity, moisture content, and pH value. The sum of the weightings of the four factors gives the overall corrosiveness index. A higher index value corresponds to a higher corrosion potential of the soil.

In the United Kingdom, the classification is largely based on the method of soil corrosivity assessment developed by Eyre & Lewis (1987). Ranking marks are used in the classification. The overall classification is determined from the sum of pertinent contributing factors, viz. soil composition, soil resistivity, moisture content, pH value, soluble salt content, etc. In the context of soil nailing, Murray (1993) classified soils, based on the ranking marks into four categories, namely "unlikely to be aggressive," "mildly aggressive," "aggressive," and "highly aggressive." Recently, the classification scheme adopted in France (see above) is also recommended in the UK (BSI, 2010, 2013).

In the United States, the soil corrosivity is assessed on the basis of the several pertinent soil properties. For example, Byrne et al. (1998) recommended that four factors be assessed, namely soil resistivity, pH value, concentration of sulfate, and concentration of chloride, whereas Lazarte et al. (2015) considered that the concentration of organic matter should also be assessed in addition to these four factors. Critical values are assigned to the four or five factors and if the actual soil property value falls below or above any one of these critical values as appropriate, the soil is classified as aggressive. Otherwise, the soil is regarded as non-aggressive.

In Hong Kong, the classification of soil aggressivity is based largely on the method developed by Eyre & Lewis (1987). It makes use of a scoring system, which categorizes the ground into four different levels of aggressiveness, namely "non-aggressive," "mildly aggressive," "aggressive," and "highly aggressive." The scoring system makes reference to the physical properties and chemical characteristics of the soils (see Tables 5.7 and 5.8). For soil nails to be installed in soils classified as "non-aggressive" or "mildly aggressive," corrosion protection includes hot-dip galvanizing and a 2 mm sacrificial thickness on the radius of the steel bar should be provided. For "aggressive" or "highly aggressive" soil, corrugated plastic sheathing in addition to hot-dip galvanization is recommended (GEO, 2008). More details are given in Section 5.5.

Table 5.7 Soil aggressivity assessment scheme in Hong Kong (GEO, 2008)

Property	Measured Value	Score
Soil Composition	Fraction passing 63 μm sieve ≤ 10%, and Plasticity Index (PI) of fraction passing 425 μm sieve < 2, and organic content < 1.0%	2
	10% < Fraction passing 63 μm sieve ≤ 75%, and fraction passing 2 μm sieve ≤ 10%, and PI of fraction passing 425 μm sieve < 6, and organic content < 1.0%	0
	Any grading, and PI of fraction passing 425 μm sieve < 15, and organic content < 1.0%	−2
	Any grading, and PI of fraction passing 425 μm sieve ≥ 15 and organic content < 1.0%	−4
	Any grading, and organic content ≥ 1.0%	−4
Resistivity (ohm-cm)	≥ 10,000	0
	< 10,000 but ≥ 3,000	−1
	< 3,000 but ≥ 1,000	−2
	< 1,000 but ≥ 100	−3
	< 100	−4
Moisture Content	≤ 20%	0
	> 20%	−1
Ground-Water Level	Above groundwater level and no periodic flow or seepage	1
	Local zones with periodic flow or seepage	−1
	At groundwater level or in zones with constant flow or seepage	−4
pH	6 ≤ pH ≤ 9	0
	5 ≤ pH < 6	−1
	4 ≤ pH < 5 or 10 ≥ pH > 9	−2
	pH < 4 or pH >10	See Note 1
Soluble Sulfate (ppm)	≤ 200	0
	> 200 but ≤ 500	−1
	> 500 but ≤ 1,000	−2
	> 1,000	−3
	(Water soluble sulfate as SO_3)	
Made Ground (see Note 2)	None	0
	Exist	−4
Chloride Ion (ppm)	≤ 100	0
	> 100 but ≤ 300	−1
	> 300 but ≤ 500	−2
	> 500	−4

Notes: (1) If pH value is less than 4 or greater than 10, the ground should be classified as aggressive regardless of the results of other tests.

(2) "Made ground" refers to man-made ground associated with high corrosion rate, such as non-engineering fill with rubbish, organic matters, etc.

Table 5.8 Classification of soil aggressivity in Hong Kong (GEO, 2008)

Classification of Soil Aggressivity	Total Scores from the Soil Aggressivity Assessment Scheme
Non-aggressive	≥ 0
Mildly aggressive	−1 to −4
Aggressive	−5 to −10
Highly aggressive	≤ −11

5.5 CORROSION PROTECTION MEASURES

5.5.1 General

There are a number of options for providing different degrees and measures of corrosion protection to steel reinforcing bars used in soil nails. Corrosion protection is basically provided through physical and/or chemical means. Electrical means, such as cathodic protection, are, however, seldom used in soil nail projects. Physical protection involves providing a continuous barrier separating the soil nail reinforcement and the associated metallic components from the corrosion sources. Chemical protection, on the other hand, consists of the use of a sacrificial material, such as galvanized material or a dielectric material like epoxy coating, which impedes the flow of electric current. The most common protection measures used in soil nails include:

- cement grout,
- sacrificial thickness to the steel,
- sacrificial metallic coating on the steel (e.g. hot-dip galvanizing),
- non-metallic coating on the steel (e.g. epoxy), and
- corrugated plastic sheath.

As each of the above corrosion protection measures has merits and limitations, it is desirable to apply a combination of these options to soil nails with due regard to site-specific conditions and availability of the measures.

5.5.2 Corrosion protection by cement grout

The most common technique for installing soil nails is by the "drill-and-grout" method. Neat cement grout can contribute to reducing corrosion by forming a physical barrier as well as a chemical barrier. The physical barrier separates the steel from direct contact with the surrounding soil. The chemical protection function of the grout is derived from its alkalinity arising from hydrated hydraulic cement and the property of steel to form a tight oxide film on its surface in an alkaline environment. This process is called passivation. As long as an effective grout cover is in place, the passivation provided by the protection of the grout prevents or delays ingress of salts, oxygen, and moisture in the soil from reaching the soil nail reinforcement. However, as the cement grout is subject to tensile stresses during its service life, micro-cracks will occur. Also, shrinkage cracks may form during the setting of cement grout. Cracking tends to be more intensive at the location where the tension is the greatest, which is approximately the position where the shearing zone intersects the soil nail. The cracks can break the physical and chemical barriers provided by the cement grout by allowing water, oxygen, and other corrosive promoting agents, especially chloride, to come into contact with the steel. There is no rational basis for predicting the rate of corrosion under such conditions. Furthermore, it is very difficult to check the quality of the grout in the ground, e.g. drillhole collapses before the grout sets can cause problems of grout integrity. It is hence widely accepted, although erring on the conservative side, that for permanent soil nails, the grout is assumed to offer no protection from the design perspective. Furthermore, the corrosion rate is taken to be the same as that for steel installed directly in the ground, in particular for nails in aggressive ground (Mitchell & Villet, 1987; Clouterre, 1991; Lazarte et al., 2015).

5.5.3 Corrosion protection by sacrificial steel thickness

This is a simple and widely used method of corrosion protection for soil reinforcement. It allows for steel corrosion by over-sizing the cross-sectional area of the steel bar. Products of corrosion that appear over time may form a protective coating between the steel and its surrounding. While this coating offers no physical protection to the steel, it may slow down the rate of corrosion by changing the kinetics of the chemical reactions. However, there are conflicting views from different researchers and practicing engineers who consider that use of thicker metal sections does not give effective protection as corrosion is rarely uniform and extends most rapidly and preferentially at localized pits or surface irregularities. Nonetheless, in the context of soil nailing, the reinforcement is commonly low carbon steel and is not under pre-stressed condition, whereas the tendons in ground anchors are usually with high carbon content and under pre-stressed condition. To a certain extent, the use of sacrificial steel thickness is still a viable corrosion protection measure for soil nails. It should, however, be noted that the products of corrosion could occupy more than several times the volume of the original steel. This exerts greater tensile stresses to the cement grout annulus, leading to further cracking. As a general rule adopted in Europe, provision of sacrificial steel thickness as corrosion protection is not suitable for soil nails with a small cross-sectional area (e.g. Eurocode EN 14490 (2010) recommends that this method should not be used if the percentage loss of the cross-section area exceeds half of its initial area), or for steel with high carbon content.

Equations (5.4) and (5.5) above can be used to estimate the average and maximum corrosion loss in thickness, X, of steel reinforcement and the sizing of the soil nail reinforcement can be done as follows:

$$d_{req} = d + 2X(\text{maximum}) \tag{5.6}$$

where d_{req} is the diameter of reinforcement required taken into corrosion, d is required diameter of reinforcement without considering corrosion, and X is corrosion loss in thickness during the service life of the soil nailed structure.

5.5.4 Corrosion protection by metallic coating

Zinc is the most common type of metal used to provide corrosion protection to steel soil nails. Zinc coating, which is a series of alloy layers on the steel surface, is often applied by the hot-dip galvanizing process. The galvanized zinc coating is strongly resistant to most corrosive environments. It also assists in preventing formation of pits in steel during the first few years of corrosion that would otherwise occur in non-galvanized steel reinforcement. As the coating is chemically integrated with the steel surface, it is more robust than other types of coating that rely on adhesion to the surface.

The primary protection provided by zinc is the formation of a corrosion-resistant coating around the steel reinforcement to isolate the steel from the surrounding environment. Zinc, being anodic to steel, also actively protects the steel cathodically by sacrificial dissolution. Thus, the zinc coating also provides cathodic protection to steel as well. This delays the onset of steel corrosion. Where there is damage or a minor discontinuity in the zinc coating, protection of the steel can still be maintained by the cathodic action of the surrounding galvanized coating. The corrosion rate of zinc is slower than that of steel, which lies between 1/17 (in rural atmospheres) and 1/80 (in marine atmospheres) to that of steel (ZALAS, 1989; Hadley & Yeomans, 1990).

The protective life of a zinc coating is roughly proportional to the mass of zinc per unit surface area. As discussed in Section 5.3, depending on the aggressivity of the soil, a nominal 85 μm (610 g/m²) coating would be capable of providing protection for 6 to 20 years in "aggressive" and "non-aggressive" soils, respectively, according to Murray (1993). Galvanized steel performs best in conditions of atmospheric exposure in near neutral pH and low chloride conditions. It is important to note that the galvanized coating corrodes rapidly in even mildly acidic conditions as might be found in peat or fills associated with colliery spoil.

5.5.5 Corrosion protection by non-metallic coating

Non-metallic coatings in the form of fusion-bonded epoxy have been in common use in the United States to protect steel soil nails from corrosion. According to Lazarte et al. (2015), the thickness of epoxy coating used in the United States varies from 7 μm to 12 μm. The epoxy coatings do not conduct electricity and they isolate the steel bars from the surrounding environment. The coating is dielectric and therefore provides physical and chemical protection to the steel reinforcement. To be effective, the coatings have to be impermeable to gases and moisture and free of gaps at the interface between the steel and the coating. Thus, the coating is commonly applied by manufacturers under factory conditions before transporting to the site. Care is necessary to ensure complete continuity of the coating. Problems have been encountered in epoxy-coated steel where severe corrosion of the steel bars occurred due to ingress of water and oxygen through cracks in the epoxy. In reviewing the historical use of epoxy-coated strand in dams in the United States, Bruce (2002) revealed that the main source of failure was attributed to the problems of epoxy adhesion to the steel.

5.5.6 Corrosion protection by encapsulation

When a high level of corrosion protection is needed, encapsulation in the form of corrugated plastic sheathing is commonly used in conjunction with cement grout. This is a technique originally developed for ground anchors in the late 1960s aiming to isolate the steel tendons from the environment. The sheathing is usually made of polyvinyl chloride (PVC) or high-density polyethylene (HDPE). To promote transfer of the force in soil nail to the surrounding grout and soil, the sheathing is corrugated and made of material with adequate strength and suitable stiffness. The steel bar is grouted inside the corrugated plastic sheath. The annulus between the sheath and the drillhole wall is also infilled with cement grout. The grouting of the steel bar in the plastic sheath can be carried out on site. Alternatively, it can be factory grouted into the sheath, which is then grouted on site into the drillhole. Some of the sheathing techniques may be proprietary products.

The inclusion of the sheath prevents ingress of water or corrosive substances if cracking of the grout annulus occurs. If coupling of encapsulated soil nail sections is required, the connection should be wrapped with heat-shrinkable tape to avoid development of a discontinuity in respect of the corrosion protection measure. Recently, a corrosion protection method employing double-corrugated sheaths has been developed. In this method, the steel bar is encased in two concentric corrugated plastic sheaths with the core and the annulus space both fully infilled with cement grout (Barley, 1997).

5.5.7 Corrosion protection to nail head and facing

The soil nail head, which commonly consists of bearing plate, nuts, and washers, is subject to different degrees of corrosion depending on whether a hard or flexible slope facing is adopted. Where a hard slope facing is used, the nail head is embedded either in concrete or shotcrete, whereas the nail head would be exposed to the atmosphere if a flexible slope facing is adopted. Although a soil nail head is not directly in contact with the ground, corrosion protection should still be provided, in particular against carbonation where carbon dioxide may react directly with the reinforcement or migrate through the cracks of the concrete/shotcrete pad to the reinforcement. Typical corrosion protection measures include sacrificial steel thickness, zinc galvanization, plastic or epoxy coating, stainless steel, and concrete cover. Occasionally, soil nail heads may be fitted with grease-filled steel caps using similar technology adopted in pre-stressed ground anchors.

For hard slope facings, corrosion protection to the reinforcement is basically provided by the concrete cover. One may also select a suitable type of cement with reference to site-specific corrosion potential of the exposed environment in order to improve the degree of protection provided by the concrete. If a steel wire mesh is applied to flexible or soft facings, zinc or plastic coating would normally be applied in order to improve their durability. Furthermore, the stability of plastic coatings under long-term ultraviolet exposure should also be considered in selecting a suitable coating.

5.6 CORROSION PROTECTION PRACTICES

In this section, the corrosion protection practices adopted in various places, including France, Germany, the UK, the United States, the Nordic countries, Japan, and Hong Kong, are briefly described.

5.6.1 Corrosion protection practice in France

The recommended corrosion protection methods for soil nailed structures in France are given in Clouterre (1991). The selection of protection measures to nail reinforcement depends on the corrosivity of the ground and the design life of the nails. The types of corrosion protection adopted mainly involve the provision of sacrificial steel thickness and plastic sheathing. Coatings such as zinc galvanizing or epoxy are seldom used in France. If such coating is provided, its corrosion protection effect is generally ignored. The design life of soil nails is divided into three categories, namely short-term (less than 1.5 years), medium-term (1.5 years to 30 years), and long-term (30 to 100 years). The requirements for corrosion protection depend on the overall corrosivity index, I, of the soil/site, and the design life of the soil nailed structure, as summarized in Table 5.9. This overall corrosivity index, I, reflecting the corrosion potential of the site and the nature of the structure, is the sum of the overall corrosiveness index A, and index C, where A depends on the corrosivity assessment of the soil, and C depends on the type of soil nailed structures (for example, $C = 0$ for non-critical structures, whereas $C \geq 2$ for critical structures in terms of failure consequence).

Table 5.9 Recommended corrosion protection measures in France (Clouterre, 1991)

Overall Corrosivity Index, I	Short-Term (< 18 months)	Medium-Term (1.5 to 30 years)	Long-Term (30 to 100 years)
≤ 4	0	2 mm	4 mm
5 to 8	0	4 mm	8 mm
9 to 12	2 mm	8 mm	Plastic sheath
≥ 13	Protective plastic sheath must be provided		

5.6.2 Corrosion protection practice in Germany

The German practice represents a comparatively conservative approach. It is designed to prevent corrosion as opposed to allowing for a sacrificial thickness of the steel reinforcement. Corrosion protection to permanent soil nails consists of encapsulating the steel bar in a corrugated plastic sheath and cement grout annulus (Glässer, 1990). Permanent soil nails are prepared under factory conditions and are typically delivered to the site in steel channels for protection against bending and cracking during transport. For temporary soil nails with a service life of less than 2 years, corrosion protection generally consists of cement grout annulus only with a minimum cover of 15 to 20 mm. In areas with aggressive ground conditions, corrugated plastic sheathing in conjunction with grout cover is used for temporary soil nails as well.

5.6.3 Corrosion protection practice in the UK

In the UK, the corrosion protection method being adopted for soil nailed structures depends largely on the corrosivity of the ground and the consequence of failure of the structure (i.e. the risk category) (Barley & Mothersille, 2005; BSI, 2013). Table 5.10 summarizes the recommended corrosion protection methods for a soil nailed structure under different categories of risk and soil corrosivity. For structures in the high-risk category, one of the following corrosion protection methods is recommended:

- Coated steel within an impermeable duct (i.e. plastic sheathing).
- Vinylester composite within an impermeable duct.
- Stainless steel within an impermeable duct.
- Uncoated steel within two concentric impermeable ducts.

5.6.4 Corrosion protection practice in the United States

Guidelines on corrosion protection to soil nails are given by of the US Department of Transportation (Byrne et al., 1998; Lazarte et al., 2015).

Table 5.10 Recommended corrosion protection measures in the UK (based on Barley & Mothersille, 2005; BSI, 2013)

| | Category of Risk | | | | | | | | |
| | Low Risk | | | Medium Risk | | | High Risk | | |
Type of Soil Nail	T or P in SCE	T in HCE	P in HCE	T or P in SCE	T in HCE	P in HCE	T or P in SCE	T in HCE	P in HCE
Steel directly in soil	R	R	NR	R	NR	NR	NR	NR	NR
Coated steel directly in soil	R	R	R	R	R	NR	NR	NR	NR
Steel surrounded by cement grout	R	R	R	R	R	NR	R	NR	NR
Self-drilled steel surrounded by cement grout	R	R	R	R	R	NR	R	NR	NR
Coated steel surrounded by cement grout	R	R	R	R	R	NR	R	NR	NR
Self-drilled coated steel surrounded by cement grout	R	R	R	R	R	NR	R	R	NR
Polyester composite surrounded by cement grout	R	R	R	R	NR	NR	R	NR	NR
Vinylester composite surrounded by cement grout	R	R	R	R	R	R	R	R	NR
Stainless steel surrounded by cement grout	R	R	R	R	R	R	R	R	NR
Self-drilled stainless steel surrounded by cement grout	R	R	R	R	R	R	R	R	NR
Steel surrounded by grouted impermeable ducting	R	R	R	R	R	R	R	R	R

(Continued)

Table 5.10 (Continued) Recommended corrosion protection measures in the UK (based on Barley & Mothersille, 2005; BSI, 2013)

	Category of Risk								
	Low Risk			Medium Risk			High Risk		
Type of Soil Nail	T or P in SCE	T in HCE	P in HCE	T or P in SCE	T in HCE	P in HCE	T or P in SCE	T in HCE	P in HCE
Coated steel surrounded by grouted impermeable ducting*	R	R	R	R	R	R	R	R	R
Stainless steel surrounded by grouted impermeable ducting*	R	R	R	R	R	R	R	R	R
Steel surrounded by pre-grouted double impermeable ducting*	R	R	R	R	R	R	R	R	R

Note::

T = Temporary (≤ 2 years)
P = Permanent (> 2 years)
SCE = Slightly corrosive environment**
HCE = Highly corrosive environment**
R = Recommended
NR = Not recommended

  * System particularly suitable for heavy or long soil nails for permanent works where one of the two protective layers may become damaged during handling or installation. This approximately equates to double corrosion protection required for permanent anchors.
  ** Defined in BS EN 14490:2010 (BSI, 2010).

Apart from ground conditions that are potentially not suitable for soil nailing (e.g. loose clean granular soils with Standard Penetration Test N values lower than about 10, soils with a relative density of less than about 30%, organic soils, etc.), the requirements for corrosion protection depend on the design life of the soil nailed structures and the aggressivity of the soil. For permanent soil nails (i.e. service life of 75 to 100 years) in non-aggressive soils, resin-bonded epoxy coating of 0.3 mm minimum thickness should be provided. In addition, a minimum grout cover of 25 mm should be provided. In aggressive soils or for those critical nailed structures, nails encapsulated inside corrugated plastic sheaths by grouting should be used. For this type of protection, the minimum grout cover between the sheath and

Table 5.11 Recommended corrosion protection measures in the United States (Byrne et al., 1998)

Type of Application	Soil Aggressivity	Recommended Corrosion Protection Method
Permanent (75 to 100 years) 75 years for permanent structures 100 years for abutments	Aggressive	Encapsulated soil nails should be used (grouting the soil nail inside a corrugated plastic sheath) Minimum grout cover between the sheath and drillhole circumference should not be less than 12 mm and that between plastic sheath and the steel bar should not be less than 5 mm
	Non-aggressive	Soil nail should be epoxy resin-bonded using an electrostatic process with a minimum coating thickness of 0.3 mm A minimum grout cover of 25 mm should be provided Centralizers should be placed at distances not exceeding 2.5 m center to center
Temporary (1.5 to 3 years)	Aggressive	Not specified
	Non-aggressive	Provision of grout around soil nail

Note: Encapsulation is also applicable to critical structures (e.g. retaining walls adjacent to lifeline and high-volume roadways), or where field observations have indicated corrosion of existing structures.

the drillhole should not be less than 12 mm. Table 5.11 shows the details of these requirements.

Recently, Lazarte et al. (2015) also suggested a procedure for selecting appropriate corrosion protection measures, namely Class A (encapsulation), Class B (epoxy or zinc galvanization coating), and Class C (sacrificial steel thickness). The selection of appropriate corrosion protection measures is based on soil corrosivity and the risk tolerance of failure of the soil nailed structure. The selection process is depicted in Table 5.12.

Table 5.12 Selection of corrosion protection measures in the United States (Lazarte et al., 2015)

Risk Tolerance Level	Soil Aggressivity Classification		
	Non-Aggressive	Aggressive	Unknown
Low	Class A	Class A	Class A
Intermediate	Class B	Class A	Class A
High	Class C	Class A	Class A

Corrosion protection measures:

Class A—Encapsulation

Class B—Epoxy or galvanization coating

Class C—Bare steel (sacrificial steel)

5.6.5 Corrosion protection practice in the Nordic countries

In the Nordic countries, guidelines for corrosion protection of soil nails are given by Rogbeck et al. (2003). The requirements of corrosion protection measures depend on the aggressivity of soil (environment) and design life of the soil nails. A scoring system is used to assess and classify the level of corrosion potential of soil. Factors to be considered in the assessment include soil type, resistivity, moisture content, salt content, pH values, layering of soil, and other factors (e.g. industrial waste, construction waste, water with salt from road). Depending on the level of corrosion potential, the ground is classified into one of the three different environmental classes, namely Environmental Class I—low potential for corrosion, Environmental Class II—normal potential for corrosion, and Environmental Class III—high potential for corrosion. Corrosion protection requirements for different environmental classes are outlined in Table 5.13. In addition to the environmental class, the consequences of failure of the nailed structure should be considered. If the consequences of failure are high, a high degree of corrosion protection may be needed for soil of Environmental Class I.

5.6.6 Corrosion protection practice in Japan

In Japan, the required corrosion protection measures depend on the corrosivity of the soil and the design life of the soil nails. According to JHPC (1998), the corrosivity of soil should be assessed and the factors to be considered are basically the same as those listed in Table 5.7. For

Table 5.13 Recommended corrosion protection measures in the Nordic countries (Modified Rogbeck et al., 2003)

Environmental Class	Design Life			
	Temporary	2–40 years	40–80 years	> 80 years
I	No	Low	Normal	Extremely high
II	No	Normal	High	Special investigation
III	Low	High	Extremely high	Special investigation

Note: No = No corrosion protection is necessary

Low = Low degree of corrosion protection, e.g. 2 mm sacrificial thickness on steel

Normal = Normal degree of corrosion protection, e.g. 4 mm sacrificial thickness on steel or grout at least 20 mm thick together with plastic barrier

High = High degree of corrosion protection, e.g. 8 mm sacrificial thickness on steel or grout at least 40 mm thick together with plastic barrier

Extremely high = Plastic barrier is necessary

Special investigation = Special investigation is required to determine suitable corrosion protection example, a soil is classified as "severe corrosive" if it has a pH value of less than 6.5 or a resistivity of less than 700 ohm-cm, or if organic or amino acid is present. For temporary soil nails (design life less than 2 years), there is no specific requirement for corrosion protection to the nails. For permanent soil nails in non-corrosive to non-severe corrosive environments, the protection measures include provision of zinc galvanization and a sacrificial steel thickness of 1 mm. In addition, a minimum cement grout cover of 10 mm is required even though it is pointed out that the provision of grout cover for corrosion protection is unreliable. If the soil is severely corrosive, steel soil nails are not encouraged and the use of high corrosion-resistant reinforcement, such as fiber reinforced plastics, is recommended.

5.6.7 Corrosion protection practice in Hong Kong

Guidelines for selecting appropriate corrosion protection measures for a soil nailed structure in Hong Kong are given by GEO (2008). In general, the aggressivity of the ground is first assessed and classified into one of the following four categories, namely "non-aggressive," "mildly aggressive," "aggressive," and "highly aggressive." The classification is based largely on the method of soil corrosivity assessment developed by Eyre & Lewis (1987). This is a comprehensive assessment method which takes into consideration most of the factors that affect underground corrosion (see Tables 5.7 and 5.8). The selection of corrosion protection measures is then based on soil aggressivity, loading condition, and design life of the soil nailed structure. For temporary soil nails with a design life of not more than 2 years, aggressivity assessment of the ground and provision of sacrificial steel thickness are not necessary, but hot-dip galvanizing is still required. For permanent soil nails carrying transient loads, the corrosion protection measures required are presented in Table 5.14. For permanent soil nails carrying sustained design loads (e.g. soil nails used to support excavations and act as "active" nails), the corrosion protection requirements are summarized in Table 5.15.

5.7 USE OF OTHER METALLIC AND NON-METALLIC MATERIALS

5.7.1 Stainless steel

Stainless steel may be used as nail reinforcement and its associated components to enhance long-term durability. In practice, stainless steel refers to a group of steel alloys that contain more than 10% of chromium. As indicated in Table 5.16, BSI (2013) broadly classifies stainless steel into three classes,

Table 5.14 Recommended corrosion protection measures for soil nails carrying transient loads in Hong Kong (GEO, 2008)

Design Life	Soil Aggressivity Classification	
	Highly Aggressive / Aggressive	Mildly Aggressive / Non-aggressive
Up to 120 years	Class 1	Class 2
Up to 2 years	Class 3	Class 3

Corrosion Protection Measures:

Class 1—Zinc galvanization of a minimum of 610 g/m^2 and corrugated plastic sheathing

Class 2—Zinc galvanization of a minimum of 610 g/m^2 and 2 mm sacrificial steel thickness

Class 3—Zinc galvanization of a minimum of 610 g/m^2

Notes: (1) For "potentially aggressive" sites without soil aggressivity assessment, Class 1 corrosion protection measures should be provided to soil nails with a design life of more than 2 years.

(2) Soil aggressivity assessment is not required for soil nails with a design life up to 2 years.

Table 5.15 Recommended corrosion protection measures for permanent soil nails carrying sustained loads in Hong Kong (GEO, 2008)

Soil Aggressivity Classification	Corrosion Protection Measure
Non-aggressive	Hot-dip galvanization and 2 mm sacrificial thickness on bar radius
Mildly aggressive, aggressive, or highly aggressive	Corrugated plastic sheath together with hot-dip galvanizing

Table 5.16 Types of stainless steel (Modified BSI, 2013)

Grade	Performance
304S31	Most common form of stainless steel for general application. Good all-round corrosion protection controlling the amount of bulk corrosion
316S31	Contains a significant amount of molybdenum (typically 2% to 3%) for improved resistance to attack from chloride and other processing chemicals; suitable for marine applications
Duplex	Provides better resistance to pitting corrosion and crevice corrosion

namely (i) 304S31, (ii) 316S31, and (iii) Duplex, to reflect their respective performances in resisting corrosion.

According to Barley & Mothersille (2005), Grade 304S31 and 316S31 stainless steels are commonly used and readily available for geotechnical applications, which offer an economic combination of performance and life cycle costing. These are available in most types of tendons and coarse threaded bars that are suitable for use in soil nails. One should note that localized corrosion is most likely to take the form of pitting or crevice corrosion in stainless steel. The risk will generally increase with increasing chloride concentration in the ground and decreasing alloy content.

Apart from providing corrosion protection, Barley & Mothersille (2005) highlighted an additional benefit of using stainless steel for soil reinforcement which is its inherent elastic in nature. In the un-processed low strength state, the elongation of stainless steel at failure can be up to 40% and even higher. For heavily cold worked high strength stainless steel, the elongation still can reach about 20% at failure. These characteristics make stainless steel particularly suitable for soil nailing in earthquake-prone regions where ductility is a crucial consideration in the design.

5.7.2 Geosynthetic materials

To overcome the problem of corrosion of metallic reinforcement, nonmetallic soil nails may be considered. There has been rapid development in geosynthetics in the last two decades. This has led to the development of reinforcement made of polyester, fiber reinforced polymers or fiber reinforced plastics (FRP), as well as other polymeric materials. Various types of geosynthetic materials, in particular FRP reinforcement, have been tried in soil nailing works.

An economic and technical appraisal of geosynthetic soil nails was carried out by Woods & Brady (1995). The study showed that there was a good potential for geosynthetic nails. However, the authors pointed out that a fundamental study of the various material interactions, in particular the bond between grout and geosynthetic reinforcement, was needed before full-scale trials were to be carried out. Turner (1999) reported on the construction of a trial soil nail wall reinforced with a patented system called "PermaNail." In this case, a soil nail consisted of one or more straps of high modulus polyester fibers which were grouted in a drillhole.

With the advance in technology, there is an increasing trend of using fiber-reinforced polymers or fiber-reinforced plastics (FRP) as an alternative to steel reinforcement in civil engineering works. These are composite materials made of fibers embedded in a polymeric resin. In general, they have good corrosion resistance in various environments. The fibers that are commonly used in composites for civil engineering works include carbon FRP (CFRP), glass FRP (GFRP), and aramid FRP (AFRP). Others that are not widely used include polyethylene, quartz, boron, ceramics, and natural fibers. Shanmuganathan (2003) gave a comprehensive review of the development and application of FRP composites in civil engineering and building structures in various countries.

CFRP reinforcement is anisotropic in nature and is characterized by high tensile strength in the direction of the reinforcing fibers. It is highly corrosion resistant and has a much better strength-to-weight ratio than steel reinforcement. However, CFRP reinforcement does not exhibit plastic yield behavior before failure as that in steel reinforcement upon loading. This necessitates consideration of its lack of ductility in geotechnical structures such as a soil nailed slope. While it has been reported that some types of CFRP reinforcement exhibit good durability under aggressive ground conditions, there is a concern over the durability of certain resin used in some CFRP reinforcement. In this connection, it is suggested that CFRP reinforcement with polyester resins should not be used because of their susceptibility to degradation in highly alkaline environments. One should follow the procedure of the standard accelerated tests for alkali resistance of the reinforcement (e.g. JSCE, 1997; ACI, 2004) to assess the performance of CFRP reinforcement under alkaline environments.

CFRP soil nails have been used in the UK to stabilize 6 m high steep slopes (Haywood, 2000). Other trial use included the installation of CFRP strips (30 mm wide × 4 mm thick × 15 m long) in Hong Kong (Cheung & Lo, 2005). The CFRP reinforcement is lightweight and as such it can greatly ease the installation works, especially at cramped slopes behind buildings (see Figure 5.6). The CFRP reinforcement has high tensile strength. Figure 4.18 in Chapter 4 compares the stress-strain behavior of a CFRP reinforcement with that of a high yield steel reinforcement. The brittle behavior and low bending capacity of CFRP are concerns that need to be carefully considered by designers. As such, CFRP appears to be not yet ready for wide and general application in soil nailing. Despite this, a set of interim design and construction guidelines has been developed in Hong Kong in order to facilitate further trial use to gain more insight and experience (Cheung & Lo, 2005).

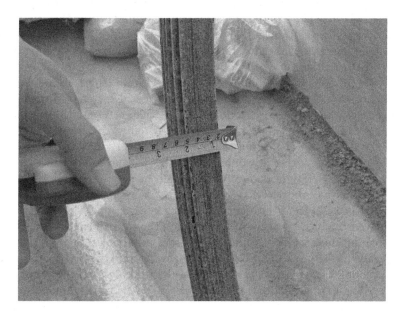

Figure 5.6 CFRP strip used for trial installation in Hong Kong.

Table 5.17 Types of glass fiber reinforcement polymers (based on Benmokrane et al., 1996; Littlejohn, 2008)

Type of GFRP	Density (g/cm³)	Tensile Strength (GPa)	Modulus (GPa)	Ultimate Strain (%)	Advantages
E-glass	2.54	3.5	72.4	4.8	Relatively low cost and readily available for general application
S-glass	2.49	4.3	86.9	5.0	Higher strength, stiffness, and ultimate strain than E-glass
C-glass	2.49	3.0	69.0	4.8	Greater chemical stability in acidic environment than E-glass
AR-glass	2.78	2.5	70.0	3.6	Alkali-resistant to minimize weight and strength loss in alkaline environment

Similar to CFRP, GFRP reinforcement relies primarily on the fiber properties for its tensile performance and the resin matrix for its lateral shear resistance. GFRP reinforcement is relatively lightweight with moderate strength but with relatively low stiffness and is used extensively in the mining industry. As indicated in Table 5.17, GFRP is classified into four types with different properties and advantages (Benmokrane et al., 1996; Littlejohn, 2008).

GFRP reinforcing bars have been used in many places of the world including Hong Kong mainly as excavation and lateral support systems. For example, to facilitate the construction of an underground station box of the Mass Transit Railway, GFRP reinforcing bars of diameters 25 mm and 40 mm were used as temporary soil nails to stabilize a 50 m high open cut in weathered granite with slope angle varying from 45° to 75° (Kwong & Chim, 2015). GFRP soil nails of 16 m to 24 m in length were installed at 2 m horizontal spacing and 1.5 m to 2 m vertical spacing. More details of this case can be found in Section 8.2.4 of Chapter 8. A clear advantage of using GFRP or other FRP bars over steel reinforcement is their light weight (a GFRP bar weighs only a quarter of its steel counterpart, with the same dimensions). Other advantages of GFRP bars include their high tensile strength, high fatigue endurance, and high corrosion resistance. As they have low thermal and electric conductivity, formation of micro-cracks is minimal at the GFRP/grout interface during cyclic temperature variations. Also, given the brittle nature of GFRP bars, the subsequent scheduled piling works in the area close to the station box could be carried out without significant difficulty. Table 5.18 summarizes a comparison of the physical and mechanical properties between the GFRP and high yield steel bars.

According to the Federal Trade Commission of the United States, aramid fibers are defined as a manufactured fiber in which the fiber forming substance is a long-chain synthetic polyamide where at least 85% of the amide

Table 5.18 Comparison of physical and mechanical properties between
GFRP and high yield steel bars (Swann et al., 2013)

Properties	GFRP Bar		High Yield Steel Bar
Diameter (mm)	25	40	25/40
Specific Gravity	2.0	2.0	7.82
Modulus of Elasticity (MPa)	40,800	40,800	210,000
Tensile Strength (MPa)	550	460	460
Shear Strength (MPa)	137	115	–
Ultimate Tensile Strain	0.0135	0.0113	0.1

(-CO-NH-) linkages are attached directly between two aromatic rings. AFRP may be used because of its high strength to weight ratio and good abrasion and impact resistance. One should, however, note that AFRP is weak in compression and known to be susceptible to creep and degradation by ultra-violet light.

Due to lack of extensive experience in using geosynthetics in soil nailing projects, care should be exercised when considering the use of these materials. For soil reinforcement, the durability of the geosynthetics may be affected by (i) degradation due to mechanical damage during site handling (e.g. abrasion and wear, punching, and tear, etc.), (ii) potential loss of strength due to creep and hydrolysis, and (iii) possible degradation due to exposure to ultraviolet radiation and heat of hydration during grouting. A thorough investigation of the material behavior needs to be undertaken, drawing the experience of their use in reinforced fill structures where appropriate.

5.8 CONCLUDING REMARKS

Soil nails are commonly installed in soils with complex physical and chemical composition, and are also subjected to variation in environmental conditions, both spatially and temporally. Potential degradation of soil nails by corrosion may adversely affect the long-term performance of a soil nailed structure. Durability is therefore an essential consideration and design requirement of a soil nailed structure. In this chapter, the fundamental mechanism of corrosion, types of corrosion, factors affecting corrosion rate, various corrosivity classification systems for the ground, and the range of associated corrosion protection measures for soil nails are presented. The corrosion potential of the ground depends on the physical and chemical composition of the soil and groundwater. Reference is made to the extensive studies carried out in the United States, the UK, and Hong Kong on corrosion of buried steel. The results of these studies indicate that pitting corrosion is more hazardous than uniform corrosion as the former has a corrosion rate which is an order of magnitude greater than the latter. In addition, the corrosion rate is extremely difficult to predict, both temporally and spatially. Depending on the corrosivity of the ground, and the nature and service life of the nailed structure, it is common practice that appropriate corrosion protection measures are provided to soil nails so that their long-term performance would not be jeopardized. The practices adopted by various places including France, Germany, the UK, the United States, the Nordic Countries, Japan, and Hong Kong have been reviewed and discussed. Common corrosion protection measures include cement grout cover, sacrificial steel thickness, galvanization, epoxy coating, plastic sheathing, and stainless steel. Apart from steel, other materials with good durability performance, such as fiber reinforced polymers (FRP), are being adopted for some soil nailing projects. Amongst these advanced materials, GFRP reinforcing bars have been used mostly as temporary soil nails for stabilizing excavations.

REFERENCES

ACI (American Concrete Institute) (2004). Guide Test Methods for Fiber-reinforced Polymers (FRPs) for Reinforcing or Strengthening Concrete Structures (ACI 440.3R-04). American Concrete Institute, Farmington Hills, MI, USA, 40 p.

Barley, A.D. (1997). Ground anchor tendon protected against corrosion and damage by double plastic layer. In Proceedings of International Conference on Ground Anchorages and Anchored Structures, Institution of Civil Engineers, Thomas Telford, London, UK, pp. 371–383.

Barley, A.D. & Mothersille, D.K.V. (2005). Durability of materials used in different environments for soil nailing. A paper contributed to the preparation of CIRIA Report No. C637 "Soil nailing – best practice guideline", UK, 54p (available on internet: www.geoserveglobal.com).

Benmokrane, B., Xu, H. & Bellavance, E. (1996). Bond strength of cement grouted glass fibre reinforced plastic (GFRP) anchor bolts. *International Journal of Rock Mechanics, Mining Sciences & Geomechanics Abstracts*, vol. 33, no. 5, pp 455–465.

Brady, K.C. & McMahon, W. (1993). The Durability of Corrugated Steel Buried Structures. Transport Research Laboratory, Crowthorne, UK, Project Report 1, 47 p.

Bruce, D.A. (2002). A historical review of the use of epoxy protected strand for pre-stressed ground anchors. In Dam Safety. Annual Conference of the Association of State Dam Safety Officials, Tampa, FL, pp. 665–678.

BSI (2010). Execution of Special Geotechnical Works: Soil Nailing (BS EN 14490: 2010). British Standards Institution, London, UK, 68 p.

BSI (2013). Code of Practice for Strengthened/Reinforced Soils: Part 2: Soil Nail Design (BS 8006-2: 2011). British Standards Institution, London, UK, 104 p.

Byrne, R.J., Cotton, D., Porterfield, J., Wolschlag, C. & Ueblacker, G. (1998b). Manual for Design and Construction Monitoring of Soil Nail Walls. Federal Highway Administration, US Department of Transportation, Washington, D.C., USA, Report No. FHWA-SA-96-069R, 530 p.

Camitz, G. & Vinka, T.G. (1989). Corrosion of steel and metal-coated steel in Swedish soils: Effects of soil parameters. In *Effects of Soil Characteristics on Corrosion*, ASTM STP1013 (edited by V. Chaker & J.D. Palmer), American Society for Testing and Materials, Philadelphia, pp. 37–56.

Cheung, W.M. & Lo, D.O.K. (2005). Use of carbon fiber reinforced polymer reinforcement in soil nailing works. In Proceedings of the HKIE Geotechnical Division 25th Annual Seminar, Hong Kong Institution of Engineers, Hong Kong, pp. 175–184.

Clouterre (1991). French National Research Project Clouterre: Recommendations Clouterre (English Translation 1993). Federal Highway Administration, US Department of Transportation, Washington, DC, USA, Report No. FHWA-SA-93-026, 321 p.

Darbin, M., Jailloux, J.M. & Montuelle, J. (1988). Durability of reinforced earth structures: The results of a long-term study conducted on galvanized steel. *Proceedings of the Institution of Civil Engineers, Part 1*, vol. 84, pp. 1029–1057.

Department of Transport (1988). *Highways and Traffic Departmental Standard BD 12*: Corrugated Steel Buried Structures. Department of Transport, London, UK.

Elias, V. & Juran, I. (1991). *Soil Nailing for Stabilization of Highway Slopes and Excavations*. US Department of Transportation, Federal Highway Administration, Washington DC, USA.

Eyre, D. & Lewis, D.A. (1987). Soil Corrosivity Assessment. Transport and Road Research Laboratory, Crowthorne, UK, Contractor Report 54.

Elias, V., Fishman, K.L., Christopher, B.R. & Berg, R.R. (2009). Corrosion/Degradation of Soil Reinforcements for Mechanically Stabilized Earth Walls and Reinforced Soil Slopes. Washington, DC, USA, Report No. FHWA-NHI-09-087.

Fontana, M.G. (1987). *Corrosion Engineering* (3rd Edition). McGraw-Hill Book Co., New York, 556 p.

GEO (Geotechnical Engineering Office) (2008). *Guide to Soil Nail Design and Construction (Geoguide 7)*. Geotechnical Engineering Office, Civil Engineering and Development Department, HKSAR Government, Hong Kong, 97 p.

Glässer, G. (1990). In-situ techniques of reinforced soil. State of the Art Lecture, In Proceedings of the International Reinforced Soil Conference, Glasgow (edited by A. McGown et al.), Thomas Telford, London, UK.

Hadley, M.B. & Yeomans, S.R. (1990). Hot dip galvanizing. *Hong Kong Engineer*, February issue, pp. 27–30.

Haywood, D.C. (2000). Sunderland scores a first. *Ground Engineering*, November issue, p. 10.

Hui, T.H.H., Tam, S.M. & Sun, H.W. (2007). Review of Incidents involving Slopes affected by Leakage of Water-carrying Services. Geotechnical Engineering Office, Civil Engineering and Development Department, HKSAR Government, Hong Kong, GEO Report No. 203, 77 p.

JHPC (Japan Highway Public Corporation) (1998). *Design and Construction Guidelines for Soil Nailing* (In Japanese). Japan Highway Public Corporation, 111 p.

JSCE (Japan Society of Civil Engineers) (1997). *Recommendations for Design and Construction of Concrete Structures Using Continuous Fiber Reinforcing Materials.* Concrete Engineering Series. Japan Society of Civil Engineers, vol. 23.

King, R.A. (1977). A Review of Soil Corrosiveness with Particular Reference to Reinforced Earth. Transport and Road Research Laboratory, Crowthorne, UK, Supplementary Report 316, 25 p.

Kwong, A.K.L. & Chim, J.S.S. (2015). Performance monitoring of a 50m high 75° cut slope reinforced with soil nail bars made from glass fiber reinforced polymer (GFRP) at Ho Man Tin Station. In Proceedings of the HKIE Geotechnical Division 35th Annual Seminar, the Hong Kong Institution of Engineers, Hong Kong, pp. 141–149.

Lazarte, C.A., Robinson, H., Gomez, J.E., Baxter, A., Cadden, A. & Berg, R. (2015). Geotechnical Engineering Circular No. 7: Soil Nail Walls: Reference Manual. Federal Highway Administration, US Department of Transportation, Washington, D.C., USA, Report No. FHWA-NHI-14-007, 385 p.

Littlejohn, A. (2008). A review of glass fiber reinforced polymer (GFRP) tendon for rock bolting in tunnels. In Proceedings of the International Conference on Ground Anchorages and Anchored Structures in Service, Thomas Telford, London, pp. 177–187.

McKittrick, D. (1978). Design, construction technology and performance of Reinforced Earth structures. In Proceedings of the Symposium on Earth Reinforcement, American Society of Civil Engineers, Pittsburgh, pp. 596–617.

Mitchell, J.K. & Villet, W.C.B. (1987). Reinforcement of Earth Slopes and Embankments. Transportation Research Board, National Research Council, Washington DC, USA, National Cooperative Highway Research Program Report 290, 323 p.

Murray, R.T. (1993). The Development of Specifications for Soil Nailing. Transport Research Laboratory, Department of Transport, UK, Research Report 380, 25 p.

Rogbeck, Y., Alen, C., Franzen, G., Kjeld, A., Oden, K., Rathmayer, H., Watn, A. & Oiseth, E. (2003). *Nordic Guidelines for Reinforced Soils and Fills.* Geotechnical Societies of Denmark, Finland, Norway and Sweden, Nordic Geosynthetic Group, Rev. B, 204 p.

Romanoff, M. (1957). Underground Corrosion, National Bureau of Standards, Circular 579, 240 p.

Shanmuganathan, S. (2003). Fiber reinforced polymer composite materials for civil and building structures: Review of the state-of-the-art. *The Structural Engineer,* vol 81, no. 13, pp 26–33.

Shiu, Y.K. & Cheung, W.M. (2003). Long-term Durability of Steel Soil Nails. Geotechnical Engineering Office, Civil Engineering and Development Department, HKSAR Government, Hong Kong, GEO Report No. 135, 65 p.

Swann, L.H., Ng, A., Mackay, A.D. & Ueda, Y. (2013). The use of glass fiber reinforced polymer bars as soil nails to permit future housing development, Hong Kong Special Administrative Region. In Proceedings of the HKIE Geotechnical Division 33rd Annual Seminar, Hong Kong Institution of Engineers, Hong Kong, pp 165–171.

Turner M.J. (1999). Trial soil nail wall using PermaNail corrosion-free soil nails. *Ground Engineering*, vol. 32, pp. 46–50.

Woods, R. & Brady, K.C. (1995). An Economic and Technical Appraisal of Geosynthetic Soil Nails. Transport Research Laboratory, Crowthorne, UK, TRL Report 129, 51 p.

ZALAS (Zinc and Lead Asian Service) (1989). *Hot Dip Galvanizing.* (4th Edition). Zinc and Lead Asian Service, 174 p.

Zhang, X.G. (1996). *Corrosion and Electrochemistry of Zinc* (1st Edition). Plenum Press, New York, 474 p.

Chapter 6

Pullout resistance

6.1 INTRODUCTION

The soil nailing technique improves the stability of slopes, retaining walls, embankments, and excavations principally through the mobilization of tension in the soil nails. The tensile forces in the soil nails reinforce the ground by directly supporting some of the applied shear loading and by increasing the normal stresses on the potential failure surface in the soil, thereby allowing higher shearing resistance of the soil to be mobilized. Whenever there is relative movement between the ground and the soil nail, shear stress will be mobilized at the interface of the soil nail surface and the ground. In order to support the entire reinforcing system, adequate resistance against pullout failure should be provided by the part of the soil nails embedded into the ground behind the potential failure surface or shearing zone, i.e. the passive zone. Hence, pullout resistance is of fundamental importance to the stability of the soil nailed structure, which is one of the key design requirements. Pullout resistance of soil nails involves the bonding mechanism at ground-grout interface and grout-reinforcement interface for grouted nails. As past experience suggests that pullout failure of soil nails is normally governed by the former, i.e. the bonding at ground-grout interface, this chapter is devoted to this mechanism. Readers may refer to other literature for more information about the bonding between the grout and nail reinforcement (e.g. GEO, 2008).

In practice, the common approach to estimate the pullout resistance of soil nails during the design stage is probably on the basis of local experience and a suitably conceptual framework. The design value of pullout resistance is then verified by field pullout tests during the construction stage. According to Hong Kong's experience, the pullout resistances as determined from field tests are usually much higher than those assumed in the design. Many researchers and designers consider that there is room for improvement in this area subject to enhanced understanding of the pertinent fundamental mechanisms. In other words, it may be possible that the number or the length of soil nails can be reduced if the pullout resistance of soil nails is estimated with better accuracy during the design stage, resulting in a more realistic cost-effective design.

As with other soil-structure interaction problems, pullout failure is a rather complicated process which involves both soil strength and stress state. To date, the mechanism of the bonding between a soil nail and its surrounding soil is still not thoroughly understood. This is mainly attributed to the complexity of the soil nail installation, ground conditions, and stress build-up and transfer process. In the case of a "drill-and-grout" soil nail, upon the formation of the drillhole, the local ground stresses are relieved and the soil at the sides of the hole will probably be disturbed to some extent. Upon insertion of the nail reinforcement, the drillhole will be grouted, probably under a certain pressure, and the local stress state of the ground will further change. During the process of curing, the grout may shrink and the stress state at the grout-soil interface will change correspondingly. Whenever the soil nail builds up stress under working condition, the load is transferred into the soil by shearing. During the shearing process, the soil may compress or dilate, and the local stresses will change again. All in all, the pullout resistance of a soil nail is a highly complicated subject and difficult to estimate in practice. Apart from the geometry of a soil nail (i.e. diameter and length), many studies indicate that the pullout resistance is affected by numerous inter-related factors including installation method, nail type, soil type, soil shear strength, surface roughness of nail-ground interface, grouting pressure, groundwater conditions, etc. (e.g. Pradhan, 2003; Cheang, 2007; Yin et al., 2009). Indeed, different design methods and safety factors are being adopted in different countries to cater for the uncertainties in estimating the pullout resistance of soil nails. As yet, there is no universally accepted method that can give the best estimation of pullout resistance for all types of soils. The best estimation of the pullout resistance of soil nails is still the value obtained from the field pullout test at the site. In this chapter, the key factors governing the pullout resistance of soil nails are identified and examined. The approaches to estimate the ultimate pullout resistance adopted by different places such as France, Germany, the UK, the Nordic countries (i.e. Norway, Sweden, Denmark, Finland, and Iceland), the United States, Japan, and Hong Kong, together with their advantages and limitations, are also discussed.

6.2 FACTORS GOVERNING PULLOUT RESISTANCE

6.2.1 General

Whenever there is small ground deformation in the active zone of a soil nailed structure, axial and lateral displacement of soil nail will be induced. The build-up of axial force in the soil nail will be limited by the shear force that can be developed at the interface between the nail and the ground (see Section 2.2 in Chapter 2 for more details about nail-ground interaction). The lateral force in the soil nail will be limited by the bearing stress between the nail and the ground. As the reinforcing action of a soil nailed structure relies primarily on the tensile force that can be mobilized in the soil nails, the ultimate pullout resistance is one of the key considerations in design. This pullout resistance relates directly to the bond strength at the interface of a soil nail and the surrounding soil. The bond strength varies along the length of the nail behind the potential failure surface, i.e. the bond length within the passive zone. Whenever the bond stress reaches the bond strength, failure will occur. To simplify the problem, the distribution of bond strength is commonly assumed to be uniform along the bond length. Thus, once the limiting value of bond stress is reached, bond failure or pullout failure will occur. In the context of soil nailing, the ultimate pullout resistance or capacity of a soil nail is defined as the maximum shear force that can be developed along the bond length. Figure 6.1 shows schematically the pullout capacity or the distribution of tensile force along a soil nail. In theory, the locus of maximum tensile force should be close to the location of the failure surface.

As highlighted in Section 6.1, the mobilization of pullout resistance is a complicated nail-soil interaction process, which is governed by a number of inter-related factors. Many researchers have endeavored to identify the key parameters that govern the ultimate pullout resistance of soil nails, T_L, with an aim to estimate its value analytically in order to facilitate soil nail design. Table 6.1 summarizes the analytical relationship between the ultimate pullout resistance and the pertinent parameters proposed by different researchers. Readers should, however, note that the actual bond stresses

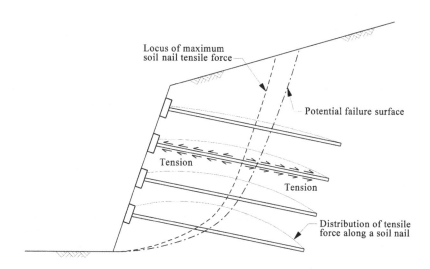

Figure 6.1 Schematic diagram showing the pullout capacity of soil nails.

Table 6.1 Analytical estimation of ultimate pullout resistance

Expression for Ultimate Pullout Resistance	Reference
$T_L = [\lambda_c\, c' + \lambda_\phi\, \sigma'_N \tan \phi']\, P_c\, L$	BSI (2013)
$T_L = [c'\, P_c + 2\, D\, \sigma'_v\, \mu^*]\, L$	GEO (2008), Schlosser & Guilloux (1981), Cartier & Gigan (1983)
$T_L = [P_c\, \sigma'_N \tan \delta]\, L$	Mecsi (1997)
$T_L = [c' + \sigma'_N \tan \phi']\, P_c\, L$	Heymann et al. (1992)
$T_L = P_c\, \sigma'_N\, f_b \tan \phi'\, L$	Jewell (1990)
$T_L = [f_c\, c' + \sigma'_N \tan(f_\phi\, \phi')]\, P_c\, L$	Potyondy (1961)

Note: c' and ϕ' are the effective cohesion and angle of shearing resistance of the soil, respectively.

μ^* and δ are the apparent coefficient of friction and the angle of friction between nail and soil, respectively.

σ'_N and σ'_v are the effective normal stress and vertical stress acting on the soil nail, respectively.

P_c, L, and D are the perimeter, bond length, and diameter of the soil nail, respectively.

$\lambda_c, \lambda_\phi, f_b, f_c,$ and f_ϕ are factors that vary from 0 to 1.

vary significantly along the bond length of a soil nail, and the distribution is normally assumed to be uniform in order to simplify the problem.

Although the various equations in Table 6.1 show some differences in detail, there are four common basic parameters that affect the estimation of the ultimate pullout resistance, namely (i) the effective normal stress acting on the soil nail surface, (ii) the apparent coefficient of friction between the nail and the soil, (iii) the adhesion between the nail and the soil, and (iv) the geometry of the nail. A more general expression for estimation of ultimate pullout resistance can be presented as follows:

$$T_L = \left[c_a + \sigma'_N \mu^* \right] P_c\, L \tag{6.1}$$

where c_a is the adhesion, σ'_N is the effective normal stress acting on the soil nail, μ^* is the apparent coefficient of friction, P_c and L are the perimeter and the bond length of the soil nail, respectively.

One should note that the adhesion between the soil nail and the soil, c_a, is usually expressed as a function of the effective cohesion, c', of the soil, i.e. $f_c\, c'$ with f_c varying from 0 to 1. According to Potyondy (1961), where a nail installed in a soil with clay content higher than 15%, the corresponding pullout resistance would be drastically reduced. As soil cohesion is usually small, some researchers have neglected this term in their proposed expressions in Table 6.1. The coefficient of apparent friction, μ^*, can be expressed as the angle of friction between the nail surface and the soil, δ, or the angle of shearing resistance of the soil, ϕ'. More discussion on this subject is given in the following.

6.2.2 Factors affecting normal stresses acting on a soil nail

One of the key parameters governing the ultimate pullout resistance of soil nails is the normal stress or radial stress acting on the nail, σ'_N. This parameter looks simple and is commonly assumed to be equal to the effective overburden pressure. It is actually a parameter that is extremely difficult to estimate accurately as it can be influenced by various inter-related factors, including soil dilatancy, soil arching, soil suction, method of nail installation, etc.

6.2.2.1 Soil dilatancy

It is well known that restrained soil dilatancy may cause an increase in normal contact stresses around the surface of soil nails as the shear stress is increased. Many studies indicate that restrained soil dilatancy in the vicinity of the soil-grout or nail-soil (driven nail) interface plays an important role in the build-up of bond strength of soil nails (e.g. Wernick, 1978; Schlosser, 1982; Palmeira & Milligan, 1989; Tei, 1993; Luo et al., 2000). When shear stain is imposed to the soils, particularly in granular soils of high density, the soil particles in the vicinity of the shear surface will slide and roll, which will result in dilation and hence the volume of the soil increases. If the tendency of dilation is partly restrained, the normal stress acting on the soil nail and hence its pullout resistance will increase. This effect is referred to as "restrained dilatancy." Experimental and instrumented field pullout tests indicate that the normal stress could increase by up to an order as compared with the initial value due to restrained soil dilation (Wernick, 1978; Guilloux et al., 1979; Xanthakos, 1991). There is, however, a limit to the effect of restrained soil dilatancy in soil nailing. If the degree of restraint keeps increasing, for example by increasing normal stress, the tendency for soil dilation and hence the further increase in normal stress will reduce correspondingly. In this case, one would expect that the increase in normal stress on a soil nail due to restrained soil will increase initially as a result of increasing overburden pressure when the confining pressure is low. When the soil's ability to restrain dilation is increased to a certain degree at a greater depth, the corresponding increase in normal stress due to restrained soil dilatancy will diminish. Thus, it is not obvious at which depth the effect of restrained dilatancy on the increase in normal stress will be the greatest. In summary, the effect of restrained soil dilatancy is controlled by the confining pressure where restrained dilatancy will be greater at low confining stresses and lower at high confining stresses. In practice, the effect of soil dilatancy is normally represented and quantified by the angle of dilatancy, which decreases with increasing confining pressure.

Apart from the overburden pressure, soil properties also play an important role in the tendency of a soil to dilate. For example, a granular soil with high density or coefficient of uniformity (well graded soil) will give a lower void ratio and hence result in a higher tendency for the soil to dilate and increase in normal stress during shearing. In contrast, the tendency of soil dilation and hence increase in normal stress for clayey soils during shearing will be comparatively low. One would expect the tendency of a soil to dilate upon shearing to depend on the compressibility of the soil and to be higher for a dense soil than for a loose soil. Heymann (1993) also suggested that the tendency of a soil to dilate would depend on the soil stiffness and particle size. Nonetheless, the study carried out by Schlosser & Juran (1979) indicated that an increase in normal stress during shearing could also be obtained even for loose sand under low overburden pressure though the magnitude is tiny.

6.2.2.2 Soil arching

Soil arching is one of the most universal phenomena encountered in soils both in the field and in the laboratory (Terzaghi, 1943). The soil arching effect is developed around a structure whenever there is a difference in stiffness between the structure and the surrounding soil. It involves a redistribution of soil forces resulting in lower stresses by which the soil load is transferred to a region of the soil mass. The soil arching effect is encountered in many geotechnical problems. For example, as reported by Robinsky & Morrison (1964), a layer of sand with high void ratio was developed close to a pile when the pile was pushed into loose and medium dense sand. This thin layer of loose sand created the soil arching effect around the pile and resulted in a reduction of normal stress on the pile.

In the context of soil nailing, when the portion of soil mass supported by a soil nail yields, the adjacent soil tends to move with respect to the remaining soil. This movement is resisted by shear stresses that reduce the pressure on the translating part of the support while raising the pressure on the adjoining rigid regions. Su et al. (2008) conducted a series of laboratory soil nail pullout tests and numerical simulations by applying overburden pressures of 40, 80, 120, 200, and 300 kPa. Both the laboratory test and numerical simulation results indicated that about 70% to 95% of the soil stresses around the drillholes were released upon completion of drilling, and that the recovery of stresses was minimal after filling up the drillholes with cement grout. In other words, the soil arching effect will significantly reduce the normal or radial stress acting on the soil nail, resulting in reduction of pullout resistance. Readers should, however, note that soil arching is a highly uncertain effect in that it may be destroyed in the long term due to changes of soil conditions surrounding a soil nail such as by wetting and drying, thus restoring either partially or totally the stresses due to overburden pressure.

6.2.2.3 Degree of saturation

It is believed that the degree of saturation or presence of matric soil suction will affect the effective normal stress and hence the bond strength at the interface between the soil nail and the soil. The basic principles of soil mechanics suggest that the effective normal stress will decrease with the increase in pore water pressure. Chu & Yin (2005) reported from the laboratory tests that the ultimate pullout resistance of soil nails under submerged condition was about half of that under natural condition. They considered that this was caused by the build-up of pore water pressure at the nail-soil interface and the reduction of soil suction. Pradhan (2003) also reported that low pullout resistance was observed for soil nails in nearly saturated soils. Su et al. (2007) presented the results of laboratory pullout tests with soils at different degrees of saturation, viz. 38%, 50%, 75%, and 98%. All the results indicated that the average pullout stress increased initially with pullout displacement until a peak value (i.e. ultimate pullout resistance) is reached, followed by a drop to the residual pullout resistance. The pullout displacement at peak pullout stress was found to be increasing with a reduction in the degree of soil saturation. Moreover, the peak pullout resistance was found to reach its highest value when the degree of saturation lay between 50% and 75%. In other words, a moderate degree of saturation of soil would benefit the pullout resistance of soil nails. Su et al. (2007) suggested that this may be due to migration of shearing plane from the interface between the nail surface and the surrounding soil into the soil mass with an increase in the degree of saturation. It is evident that the degree of soil saturation can play an important role in the ultimate pullout resistance and more studies in this area are necessary to better understand the fundamentals and mechanics behind this observation.

6.2.2.4 Soil nail installation method

According to Schlosser (1982), the normal stress on a soil nail will be close to overburden pressure if it is driven into granular soils. If the soil nail is driven into the ground without pre-drilling, the soil nail will displace the soil equivalent to its own volume and hence the normal stress acting on the soil nail will tend to increase (Johnson & Card, 1998). This suggests that a driven soil nail with a greater volume should have a somewhat greater unit pullout resistance than another similar nail with a smaller volume due to its higher ability to increase in normal stress by soil displacement.

Unlike driven soil nails, many researchers consider that there is no apparent relationship between overburden pressure and the normal stress acting on "drill-and-grout" soil nails, in particular when they are installed at great depths (e.g. Byrne et al., 1998; GEO, 2008; Su et al., 2008). It is believed that the soil stress acting on the drillhole surface is released upon drilling and the stability of the drillhole is maintained by the arching effect as described before. As the stress recovery after grouting is usually small, the normal stress acting on a soil nail is mostly induced by the restrained soil dilatancy effect during pullout. Nonetheless, this also depends on whether the arching effect will diminish with time. More discussion on the relationship between overburden pressure and pullout resistance is given in Section 6.2.4.

For "drill-and-grout" soil nails, many study findings indicate that grouting pressure has some effects on the normal stress acting on a soil nail. For example, Plumelle et al. (1990) considered that the normal stress on a soil nail is only about 10% to 15% of the grouting pressure. Winterkorn & Pamukcu (1991) compared the effect of different grouting techniques on the pullout resistance of soil nails and found that the normal stress for soil nails installed by pressurized grouting is higher than that installed by slurry or gravity grouting. Yin et al. (2009) also demonstrated through laboratory pullout tests that pressurized grouting could increase the normal stress to a certain extent although the main contributory factor to the increase in normal stress is still the restrained soil dilatancy. Elias & Juran (1991) proposed an empirical relationship between grouting pressure and normal stress acting on a soil nail as follows:

$$\sigma'_N = a\,p \tag{6.2}$$

where a is an empirical coefficient and p is the grouting pressure.

Because of the difficulties in obtaining the grouting pressure and the empirical coefficient, the above equation is seldom used in practice.

Apart from the grouting pressure, the geometry of the drillhole can also influence the normal stress acting on a soil nail. This relates to the drill bit and the type of grout. Theoretically, for a cylindrical drillhole with a smooth surface, the normal stress will be equal to the stress prevailing during drilling of the hole, i.e. close to zero, and hence the resulting normal stress will be very low. In practice, the surface of the drillhole is usually rough which facilitates the effect of restrained soil dilatancy, and hence giving rise to a greater normal stress. Moreover, a rough interface between grout and soil also provides a mechanical interlocking effect and causes an increase in normal stress during pullout (Johnson & Card, 1998).

6.2.3 Factors affecting apparent coefficient of friction

Depending on the roughness of the soil nail surface, the shearing plane could occur either at the nail-soil interface or entirely within the soil just outside the nail. In other words, the apparent coefficient of friction, μ^*, lies between the angle of friction of the nail-soil interface, δ, and the angle of shearing resistance of the soil, ϕ'. For a soil nail with rough surface, such as a driven nail with ribs or a "drill-and-grout" nail with irregular surface texture, failure tends to occur within the soil. In this case, the angle of shearing resistance of the soil, ϕ', will govern the apparent coefficient of friction. If the soil nail profile is smooth, failure will take place at the nail-soil interface instead and the angle of nail-soil interface friction, δ, will govern the apparent coefficient of friction.

6.2.3.1 Soil properties

It is envisaged that the apparent coefficient of friction between soil nail and the surrounding soil, and hence the pullout resistance, depends on the soil type. For example, Bruce & Jewell (1987) reported that for the same type of soil nails installed in silty clay, sand, and sandy gravel, the corresponding pullout resistance of about 40–80 kPa, 100 kPa, and 200 kPa were obtained. Franzen (1998) suggested that the apparent coefficient of friction depends on the mineral content, shape and size of soil particles, coefficient of uniformity, and relative density of the soil.

If shear failure occurs within the soil, the apparent friction will be governed by the angle of shearing resistance of the soil, ϕ', and hence the properties of the soil. For soils with a relatively high clay content, the tendency of soil dilatancy and hence the apparent friction will decrease. Potyondy (1961) reported that when clay content was higher than 15%, the pullout resistance would be reduced drastically. Schlosser (1982) suggested that the pullout resistance for a soil nail in saturated silt or clay would depend mainly on the undrained shear strength of the soil and would be relatively small. The coefficient of uniformity of a soil also gives an indication about the angle of internal friction of the soil (Franzen, 1998). For well graded soils, the number of contact points between the soil particles would be higher than for a poorly graded soil, and the tendency for grain crushing would be less. Furthermore, the void ratio of the well graded soils would be lower and hence their tendency to dilate would be higher. This suggests that an increase of the coefficient of uniformity would result in increase of the apparent coefficient of friction of the soil. The relative density is also a good indicator of the tendency of a soil to dilate, and hence the apparent friction. A high relative density would result in a high tendency to dilate, hence resulting in an increase of apparent coefficient of friction of the soil.

In summary, pertinent soil properties affecting the apparent friction include the mineral content, size and shape of soil grains, coefficient of uniformity, and relative density.

6.2.3.2 Soil nail properties

If the available strength at the nail-soil interface is fully mobilized, the angle of interface friction between the nail and the soil, δ, will then govern the apparent friction. Numerous studies have been carried out to

Table 6.2 Angle of interface friction for different materials and surface roughness (Modified Potyondy, 1961)

| Soil Type | Angle of Interface Friction, δ(Degree) | | | |
| | Steel | | Concrete | |
	Smooth	Rough	Smooth	Rough
Sand (Dry, $\phi' = 44.5°$)	24.7	34.0	39.5	44.0
Sand (Saturated, $\phi' = 39°$)	24.8	-	34.7	-
Silt (Dry, $\phi' = 40°$)	31.5	39.8	39.8	40.0
Silt (Saturated, $\phi = 29.8°$)	20.2	-	29.8	-
Cohesive granular soil (w = 13%, $\phi' = 22°$, $c' = 44$ kPa)	9.8	18.5	19.0	20.5
Cohesive granular soil (w = 17%, $\phi' = 13°$, $c' = 18$ kPa)	7.5	10.0	13.0	13.5

Note: w, ϕ' and c' are the water content, angle of shearing resistance, and effective cohesion of the soil.

determine the friction between different construction materials and soil (e.g. Potyondy, 1961; Koivumaki, 1983; Pedley, 1990). In general, the studies indicate that the types of construction materials and their surface roughness have a significant effect on the interface friction with soil. Table 6.2 summarizes some of the findings obtained by Potyondy (1961) through direct shear box tests.

One may note that it is not uncommon in practice that the apparent coefficient of friction, μ^*, is greater than either tan δ or tan ϕ'. It is envisaged that the angle of apparent friction also depends on the restrained soil dilatancy. However, many studies also suggest that this dependence will diminish when the confining stress is increased (e.g. Schlosser & Elias, 1978; Cartier & Gigan, 1983). If the confining stress is very high, the soil particles may suffer grain crushing before they roll and override their neighbors, and the soil volume may contract rather than dilate. In this case, the dilation effect on the apparent coefficient of friction may vanish completely and the apparent friction may approach to the true friction. The interaction mechanism between the apparent friction and dilatancy is complicated. Nonetheless, it is believed that the apparent friction is intrinsically related to soil dilatancy, which is influenced by many factors including soil density, size and shape of soil particles, surface roughness, and the confining stress on soil nail.

6.2.4 Effect of overburden pressure

The effect of overburden pressure on the bond strength or pullout resistance of a soil nail is not fully understood. There have been conflicting views in the literature about the effect of overburden pressure on the pullout resistance of soil nails. For driven soil nails, it is generally believed that the pullout resistance increases as the soil depth or overburden pressure increases (e.g. Schlosser et al., 1983; Jewell, 1990). However, for "drill-and-grout" soil nails, while some studies suggest that the pullout resistance increases with increasing overburden pressure (e.g. Pradhan et al., 2006), others observed from field test results suggest that they are independent (e.g. Heymann et al., 1992; Franzen, 1998). As mentioned in Section 6.2.2, Su et al. (2008) conducted a systematic laboratory and numerical study on pullout resistance of soil nails and reported that the applied overburden pressure was released by 70% to 95% upon completion of drillhole formation, and no apparent relationship between pullout resistance and overburden pressure was observed. They believed that the development of pullout resistance was mainly attributed to the effect of restrained soil dilatancy.

Cheung & Shum (2012) collected and analyzed some 900 field pullout test results in which more than 80% of the test nails were installed in weathered granitic or volcanic rocks in Hong Kong. The rest of the test nails were installed in other materials including fill, colluvium, and moderately decomposed rock. All the test nails were installed using the "drill-and-grout" method and the overburden pressure at the bond sections varied from 50 kPa to 400 kPa. The pullout resistances of the test nails were estimated using the equation in Table 6.1 (GEO, 2008). The estimated ultimate pullout resistance was less than 50% of the yield strength of the steel reinforcement used in the tests. In theory, pullout failure should occur without the test load exceeding 50% of the yield strength of the steel reinforcement. Field pullout tests were carried out (see Section 6.4 for more details) and the results indicate, however, that most of the pullout tests (about 87%) were tested up to 90% of the yield strength of the steel reinforcement without pullout failure. The tests were stopped at this stress level in order to avoid the risk of rupture of the reinforcement. In other words, the actual ultimate pullout resistances in most cases were much higher than the estimated values. For the remaining 13% (about 120 test nails), although the test loads reached the actual pullout capacity, they were still well above the estimated values. Figure 6.2 shows the ultimate pullout resistance plotted against the corresponding effective overburden pressure of these 120 test nails in which their actual pullout capacity was reached. No apparent correlation between the ultimate pullout resistance and the overburden pressure can be established from the plot. It is believed that the independence of overburden pressure to the ultimate pullout resistance is related to the soil arching effect in the drillhole. If the soil arching effect diminishes with time, the dependence of pullout resistance to overburden pressure may resume. Another finding of the analysis is that the field pullout resistance is generally higher than the theoretical values given by the effective stress method currently adopted in Hong Kong. More discussion on different approaches in estimation of the ultimate pullout resistance is given in Section 6.3.

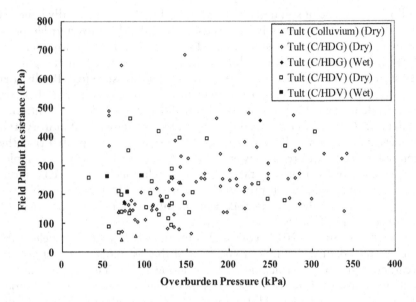

Figure 6.2 Plot of ultimate pullout resistance against overburden pressure.

6.3 ESTIMATION OF ULTIMATE PULLOUT RESISTANCE

Various theoretical and empirical methods have been proposed in the literature for estimating the ultimate pullout resistance of soil nails (e.g. Schlosser, 1982; Cartier & Gigan, 1983; Jewell, 1990; Heymann et al., 1992; Watkins & Powell, 1992). In practice, the pullout resistance is commonly estimated from published analytical methods, correlation with parameters obtained from site investigation, or directly from field soil nail pullout tests. There is no unified method for estimating the ultimate pullout resistance as reflected by different technical standards and codes of practice adopted by different places in the world. For instance, the effective stress method is commonly used in Hong Kong (GEO, 2008), correlation with soil type is used in the United States (Elias & Juran, 1991; Byrne et al., 1998; Lazarte et al., 2015), correlation with pressuremeter test results is used in France (Clouterre, 1991), and correlation with SPT N value is used in Japan (JHPC, 1998) and the United States (Sabatini et al., 1999). The various methods for estimating pullout resistance can be classified into three major categories, namely (i) empirical correlations, (ii) site-specific pullout tests, and (iii) analytical methods.

6.3.1 Empirical correlations

As the pullout resistance of soil nails is affected by various factors which are subject to different degrees of variability, it is convenient that the ultimate pullout resistance is estimated based on empirical correlations with selected pertinent key parameters. Depending on the reliability of the empirical correlation and the quality of site-specific field data, this approach is mostly used for preliminary design. The design value is normally verified subsequently by the appropriate field pullout test on representative soil nails installed on the site concerned.

6.3.1.1 Correlation with soil/rock type and installation method

It is common in the United States that the bond strength between soil nails and the ground is estimated empirically based on precedent experience. For example, Elias & Juran (1991) summarized the typical values of ultimate bond for the "drill-and-grout" soil nails installed in different ground conditions by a variety of drilling methods (Table 6.3). The lower and upper bounds of the values correspond to the least and most favorable conditions for a particular soil type and construction method are given for reference. At present, these typical values are recommended by published standard and code of practice in the United States (Byrne et al., 1998; Lazarte et al., 2015).

One should note that the values recommended in Table 6.3 are not exhaustive and they correspond to cases with gravity grouting only. The bond strength may increase significantly by adopting pressurized grouting during nail installation. In practice, it is common that the *in situ* pullout test is carried out to verify the estimated bond strength before production or working nails are installed.

6.3.1.2 Correlation with pressuremeter test

It is also quite common in France and the United States that the bond strength of soil nails is estimated based on site-specific pressuremeter test results. For example, Elias & Juran (1991) presented the empirical correlation between the measured pullout resistance of both driven and "drill-and-grout" soil nails, and the estimated pullout resistance based on the pressuremeter test (Figure 6.3). Design charts and equations correlating ultimate bond stress of nail in soil with pressuremeter limit pressure are also given in design guides (e.g. Clouterre, 1991; Lazarte et al., 2015). Similar to other empirical correlation methods, the large range of factors influencing the load transfer mechanism of pullout may affect the accuracy of the estimation. The corresponding codes of practice recommend that the estimation should only be used for preliminary design. Verification by the field pullout test during nail installation is necessary.

6.3.1.3 Correlation with SPT N value

Another common approach for determining pullout resistance of soil nails is by means of empirical correlation with SPT N values. For example, it is

Table 6.3 Typical bond strength of soil nails in different soils and rocks (Modified Elias & Juran, 1991)

Ground Material	Soil/Rock Type	Installation Method	Ultimate Bond Stress, psf (kPa)
Rock	Marl/Limestone	Rotary Drilled	6,000–8,000 (290–380)
	Phyllite		2,000–6,000 (100–290)
	Chalk		10,000–12,000 (480–570)
	Soft Dolomite		8,000–12,000 (380–570)
	Fissured Dolomite		12,000–20,000 (570–960)
	Weathered Sandstone		4,000–6,000 (190–290)
	Weathered Shale		2,000–3,000 (100–140)
	Weathered Schist		2,000–3,500 (100–170)
	Basalt		10,000–12,000 (480–570)
Cohesionless Soils	Silty Sand	Rotary Drilled	2,000–4,000 (100–190)
	Silt		1,200–1,600 (60–80)
	Piedmont Residual		1,500–2,500 (70–120)
	Fine Colluvium		1,500–3,000 (70–140)
	Coarse Colluvium		2,000 (100)
	Sand/Gravel	Rotary Drilled (Wet)	6,000–9,000 (290–430)
	Sand	Driven Casing	6,000 (290)
	Sand/Gravel		4,000–5,000 (190–240)
	low overburden		6,000–9,000 (290–430)
	high overburden		8,000–12,000 (380–570)
	Dense Moraine Colluvium		2,000–4,000 (100–190)
	Silty Sand Fill	Augered	400–600 (20–30)
	Silty Fine Sand		1,700–2,200 (80–110)
	Silty Clayey Sand		2,500–5,000 (120–240)
	Sand	Jet Grouted	8,000 (380)
	Sand/Gravel		20,000 (960)
Cohesive Soils	Silty Clay	Rotary Drilled	700–950 (35–50)
	Clayey Silt	Driven Casing	1,800–3,000 (90–140)
	Silty Clay		3,600 (170)
	Loess	Augered	500–1,500 (25–70)
	Soft Clay		400–600 (20–30)
	Stiff Clay		800–1,200 (40–60)
	Clayey Silt		800–2,000 (40–100)
	Calcareous Sandy Clay		4,000–6,000 (190–290)

common practice in Japan and the United States that the pullout resistance of soil nails is determined during design stage based on SPT N values of the ground. Tables 6.4 and 6.5 summarize the recommended pullout resistance for different materials and SPT N values (JHPC, 1987; Sabatini et al., 1999).

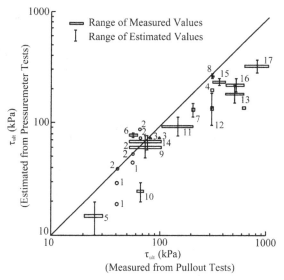

Note :

1. Driven bars in fine grained soil
2. Grouted bars in fine grained soil
3. Driven bars in granular soil
4. Grouted bars in weathered rock
5. Grouted bars in soft clay
6. Grouted bars in stiff clay
7. Grouted bars in marl
8. Grouted bars in stiff marl
9. Grouted bars in clayey silt

10. Drilled & Grouted bars in silt
11. Drilled & Grouted bars in silty sand
12. Driven casing grouted bars in sand
13. Driven casing grouted bars in moraine
14. Driven casing grouted bars in colluvium
15. Drilled & Grouted bars in mari-limestone
16. Drilled & Grouted bars in soft rock
17. Drilled & Grouted bars in fissured rock

Figure 6.3 Correlation between pullout resistance of driven/grouted soil nails and pressuremeter test. (Modified Elias & Juran, 1991.)

6.3.2 Site-specific pullout test

The site-specific pullout test measures the pullout resistance of soil nails directly in the field. Appropriate pullout resistance for design of soil nails can be determined from the measured values. This is the most accurate method for determining pullout resistance of a specific site. However, it may not be always possible to conduct the site-specific pullout test during the site investigation stage to determine the design pullout resistance of soil nails. The pullout test is recommended by various places in the world including Japan, the United States, and many European countries (JHPC, 1987; BSI, 2013; Lazarte et al., 2015). Readers may refer to Section 6.4 for more details about the site-specific pullout test.

Table 6.4 Estimated ultimate bond stress of soil nails in different materials and SPT *N* values (Modified JHPC, 1987)

Ground Material		SPT *N* Value	Ultimate Bond Stress, (kPa)
Rock	Hard Rock	-	1,500
	Soft Rock		1,000
	Weathered Rock		600
	Mudstone		600
Gravel		10	100
		20	170
		30	250
		40	350
		50	450
Sand		10	100
		20	180
		30	230
		40	290
		50	300
Cohesive Soils			Effective Cohesion, c'

Table 6.5 Ultimate bond stress of soil nails in different materials and SPT *N* values (Sabatini et al., 1999)

Ground Material		SPT *N* Value	Ultimate Bond Stress, (kN/m)*
Gravel and Sand	Loose	4–10	145
	Medium dense	11–30	220
	Dense	31–50	290
Sand	Loose	4–10	100
	Medium dense	11–30	145
	Dense	31–50	190
Sand and Silt	Loose	4–10	70
	Medium dense	11–30	100
	Dense	31–50	130
Silty-clay Mixture with Low Plasticity or Fine Micaceous Sand or Silt Mixture	Stiff	10–20	30
	Hard	21–40	60

Note: * Ultimate bond stress is represented as force per unit length of bond length.

6.3.3 Analytical method

With increasing knowledge about the factors affecting pullout resistance, it is quite common to employ analytical methods to determine the ultimate pullout resistance of soil nails. One may use the effective stress method to estimate the pullout resistance in granular soils and the total stress method in cohesive soils.

6.3.3.1 Effective stress method

The pullout resistance of soil nails in granular and cohesive soils under long-term conditions can be calculated analytically using the effective stress design method. This method is commonly adopted in the UK and Hong Kong for design of soil nails. For example, the following expression is used in Hong Kong to estimate the ultimate pullout resistance provided by the soil-grout bond strength in the passive zone of "drill and grout" soil nails, T_L:

$$T_L = \left[c'P_c + 2D\sigma'_v \mu^* \right] L \tag{6.3}$$

where c' is effective cohesion of the soil, μ^* is the apparent coefficient of friction of the soil nail (which is taken as tan ϕ', where ϕ' is the angle of shearing resistance of the soil under effective stress condition), σ'_v is the vertical effective stress in the soil calculated at mid-depth of soil nail in the passive zone, L is the bond length of soil nail within the passive zone, and D *and* P_c are diameter and perimeter of the grout annulus, respectively.

It should be noted that like other methods, the effective stress method has some limitations and that the pullout resistance of a "drill-and-grout" soil nail assessed by this method is only an estimate based on simplified assumptions. The effective stress method does not account directly for factors such as soil arching, restrained soil dilatancy, soil suction, roughness of drillhole surface, and over-break if any. Nevertheless, based on the extensive experience in Hong Kong, use of the method, together with an appropriate safety margin, would give an adequately safe design for most ground and groundwater conditions commonly encountered locally. Furthermore, according to the published literature and field pullout test results, the effective stress method tends to underestimate the pullout resistance in granular soils primarily because the contributions of restrained soil dilatancy have been ignored, which is one of the major factors affecting the normal stress acting on the soil nail during pullout (Clouterre, 1991; Phear et al., 2005; BSI, 2013), see also Section 6.2.4.

6.3.3.2 Total stress method

The pullout resistance of soil nails in stiff cohesive soils can be estimated using the total stress method, i.e. applying the undrained shear strength, c_u, with an adhesion factor, α, in a similar way to that for assessing pile capacity in clays. The value of α depends on the condition at the grout-ground or reinforcement-ground interface, which varies typically from 0.3 to 0.9. In general, the pullout resistance estimated by this method is higher than that determined by the effective stress method. When the total stress method is employed for estimating ultimate pullout resistance in high plasticity soils,

Table 6.6 Merits and limitations of different methods for estimating ultimate pullout resistance of soil nails

Method	Merits	Limitations
Empirical Correlations	• Relate to field performance data • Can better account for the influencing factors	Need a substantial number of field data and take a long time to establish a reasonable correlation A general correlation may not be applicable to all sites
Field Pullout Tests	• Relate to site-specific performance data	Need to conduct a sufficient number of field pullout tests to obtain reliable and representative design value May not be feasible for small-scale projects Time consuming
Analytical Methods	• Based on soil mechanics principle • Easy to apply	Many factors affecting pullout resistance are not directly and explicitly accounted for

a high degree of conservatism should be allowed to cater for the possible large difference between the peak and residual shear strength of the soils.

In summary, there are three approaches, namely empirical correlations, field pullout tests (i.e. design by testing), and analytical methods, that are being adopted by various places to determine the ultimate pullout resistance of soil nails during the design stage. Table 6.6 summarizes the merits and limitations of the three approaches. Readers should select one, or a combination of the three, that best suits their sites and purposes.

6.4 FIELD PULLOUT TEST

6.4.1 General

The pullout resistance is one of the key parameters that affects the design and performance of a soil nailed structure. As presented in previous sections, the pullout resistance of soil nails depends on a number of factors, which are inter-related and highly uncertain. This suggests that the pullout resistance cannot be estimated with a high degree of certainty. Because of this, it is fairly common that soil nails are load tested on site during the investigation stage or construction stage for different purposes. For the former, the main objective of the test is to obtain information on pullout resistance at the early stage of a project for subsequent design, while for the latter, the tests are carried out during the construction stage to verify design assumptions, check the contractor's workmanship in respect of drilling, installation and grouting operation, as well as to ascertain the performance of the soil nails (e.g. without excessive deformation under working load condition, etc.). To suit different purposes, field pullout tests can be classified broadly into four categories, namely (i) investigation pullout test, (ii) verification (or suitability) pullout test, (iii) proof (or acceptance) pullout test, and (iv) creep pullout test.

6.4.2 Investigation pullout test

The primary objective of an investigation pullout test is to determine the ultimate pullout resistance during the investigation stage for subsequent soil nail design. It is a cost-effective approach to obtain design information on site-specific pullout resistance for large-scale projects. In general, the number of investigation pullout tests depends on the relevant soil strata to which production soil nails are to be bonded, the variability of ground condition, and the likely number of production nails. Depending on the required accuracy of design information and other project requirements, one should exercise engineering judgment to determine the locations of the tests and the number of test nails needed.

The test is normally carried out on sacrificial nails during the investigation stage (i.e. before design stage). A sacrificial soil nail is a nail installed solely for the purpose of the pullout test, which will not form part of the permanent works. These test nails should be installed using the same procedures as the production nails except that only the bottom part of the soil nail at the specific stratum, where the bond information is required, is to be grouted. This is to ensure that the test nails are representative of the production nails in terms of construction and performance. The test usually comprises a single cycle test in which the load is applied in increments to a maximum test load, which typically corresponds to the yield strength of the reinforcement or when failure occurs at the grout-ground or reinforcement-ground interface whichever is the less. To ensure that failure will occur at the grout-ground (for "drill-and-grout" nails) or reinforcement-ground (for driven nails) interface instead of overstressing the reinforcement, the reinforcement provided for the test is usually oversized. In practice, it is not always possible to install test nails during the investigation stage, especially where the mobilization cost is relatively high as compared with the total site investigation cost. In most cases, it is only an option for designers' consideration.

6.4.3 Verification pullout test

The primary objective of a field verification (or suitability) pullout test is to verify the design assumptions on the bond strength at the nail-ground interface. The test also gives an indication of the contractor's workmanship, the appropriateness of the construction method given the specific ground and groundwater conditions, and potential construction difficulties. As a minimum, the verification pullout test should be carried out to a test load defined by the design working load times the minimum required safety factor against pullout failure. In general, field verification pullout testing is carried out during the construction stage to obtain the following information:

- The bond strength at the interface between the soil nail and the ground if the test is carried out to pullout failure. However, in most cases, the test is terminated when the test load reaches the yield strength of the reinforcement in order to avoid the possibility of a brittle failure. The reinforcement can be oversized to ensure that failure will occur at the nail-ground interface instead of overstressing the reinforcement.
- Verification of the design safety margin against pullout failure.
- Deformation behavior of the soil nail under tensile loading.
- Appropriateness of the construction method and potential construction difficulties.
- Workmanship of the soil nailing works.

Verification pullout tests should, as far as practicable, be carried out at locations where the pullout resistance may be low or where the buildability of the soil nails is most uncertain, e.g. at locations of relatively weaker material or locations with high groundwater level or significant local seepage. Similar to an investigation pullout test, a verification pullout test should be carried out on sacrificial nails and prior to the construction of production soil nails so that the information gathered from the tests can be reviewed for making design changes as necessary.

The number of verification pullout tests should be defined based on the confidence level required for the design, the variability of the ground and groundwater conditions, and loading conditions during the design life of the soil nailed structure. It is common practice to prescribe the number of verification pullout tests as a certain percentage (e.g. 2% in Hong Kong and 5% in the United States) of the total number of production soil nails. However, one should exercise engineering judgment in ensuring that the number of tests is sufficient to meet the test objectives.

6.4.4 Field acceptance pullout test

The primary objectives of a field acceptance or proof pullout test are to ascertain the consistency of the contractor's workmanship and verify the performance soil nails under working load condition during the course of, or after, construction. According to Lazarte et al. (2015), the proof test nail should have a bonded length and a temporary unbonded length. Fully grouted nails should not be used for the acceptance test, otherwise the results may be misleading because part of the pullout resistance is attributed to the ground-grout bond in the active zone. The test is carried out on production or working nails, it should be cautioned that the integrity of the soil nail (e.g. grout annulus and corrosion protection measures) may be damaged during a pullout test. Because of these concerns, the field acceptance test on working or production soil nails is not encouraged in Hong Kong's practice and any such tested nails should be taken as sacrificial nails. Nevertheless, the test is carried out on a certain percentage of production nails in some places (e.g. up to 5% in the United States). The test comprises a single cycle test in which the load is applied in increments to a maximum test load, usually by a small margin less than the design working load (e.g. 75% of design working load in the United States practice).

6.4.5 Field creep pullout test

For soil nails designed to carry sustained loads and bonded in soil, a field creep pullout test should be carried out to assess the susceptibility of the nails to long-term creep. The creep pullout test is typically carried out as part of the investigation, verification, or acceptance pullout test. The test is conducted at a specified and constant sustained test load with deformation recorded at specific time intervals, typically from 10 to 60 minutes. The test is carried out to ensure that the soil nail can carry the intended load safely throughout the service life with low or predictable creep displacement under sustained load.

In summary, various types of field pullout tests can be undertaken to meet different objectives, including obtaining information on the ultimate pullout resistance, assessing the appropriateness of the proposed construction method, identifying any possible construction difficulties and workmanship problems, and investigating possible soil creep effect. Table 6.7 gives a summary of the purposes and details of different types of field pullout tests.

6.4.6 Set-up of field pullout test

The test nails should be installed using the same procedures as the production nails. For the investigation or verification test, only the bottom part of the soil nail is grouted. Designers should design the length of the grouted part to suit the test objectives. In Hong Kong, the length of the grouted part of test nail is typically 2 m. Too short a bond length may not be representative, whereas a large bond length requires a sizable pullout load and hence heavy equipment and more complicated set-up. Grouting should be carried out slowly and carefully to prevent over-grouting. Packers are usually used to seal off the grouted section. One may use the time domain reflectometry (TDR) technique as an alternative to the use of packer to determine the length of the grouted section (see Section 7.10.4 for more details). Many types of packers, such as inflatable packers, are available in the market. Only packers that can effectively seal off the grouted section should be used. The packers should, as far as practicable, not contribute to the bond strength of the grouted section, or otherwise the contribution should be taken into account in the estimation of bond strength. It is good practice to collect additional information, such as material type and presence of groundwater, during hole drilling in order to confirm or review the geological model.

Table 6.7 Summary of different types of pullout tests

| | Types of Soil Nail Load Test | | |
	Investigation Test	Verification Test	Acceptance Test
Objective	To obtain information on the ultimate pullout resistance	• To verify design assumptions • To assess suitability of construction method and workmanship, and identify potential site difficulties	To assess the consistency of workmanship and construction method, and performance of soil nails under working load condition
When tested	Investigation Stage	Construction Stage	Construction Stage
Type of soil nail tested	Sacrificial	Sacrificial	Sacrificial or Production
Frequency	• Depends on the relevant soil strata to which production soil nails are to be bonded • Depends on the required information accuracy	A few percent (e.g. 2% in Hong Kong and 5% in the United States)	A few percent (e.g. 2–3% in the UK and 5% in the United States)
Remarks	• Not always possible or desirable to carry out during investigation stage • Potentially high mobilization cost	Not always possible to carry out the test on the structure	• May only test the pullout resistance from active zone • May overstress the reinforcement-grout bond • May damage corrosion protection

Note: Creep pullout is typically carried out as part of the investigation, verification, or acceptance pullout test, in which deformation under specified sustained load is recorded at defined time interval.

Figure 6.4 shows a typical setup of a pullout test. When setting up the test apparatus, the steel seating plate to be used for the test should not be allowed to bear onto the nail reinforcement, as this will deflect the reinforcement thereby giving incorrect readings during the test. A center-hole hydraulic jack and hydraulic pump are commonly used to apply the test load to the nail reinforcement. The axis of the jack and the axis of the reinforcement must be aligned to ensure uniform loading. The seating plate for the jack should preferably be fixed on a concrete pad resting on the trimmed soil or shotcrete surface for load spreading. In this case, any deformation of the underlying soil would be more uniform because of the rigidity of the concrete pad and overstressing of the bearing surface would also be

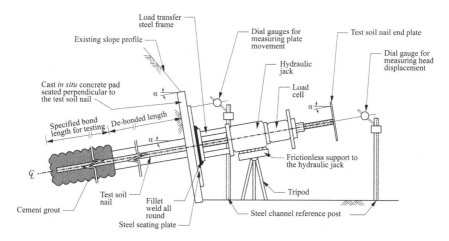

Figure 6.4 Typical setup for a pullout test in Hong Kong (GEO, 2008).

avoided. A frictionless support should be provided to the test apparatus in order to minimize the friction loss due to jacking motion.

Figures 6.5 and 6.6 present the typical testing procedures and acceptance criteria of pullout test and creep test, respectively. Designers should refer to the relevant documents published in various places for guidance on testing procedures and acceptance criteria (e.g. GEO (2008) for Hong Kong; BSI (2010) for the UK; Lazarte et al. (2015) for the United States).

6.5 CONCLUDING REMARKS

The soil nailing technique improves the stability of an earth structure principally through the mobilization of tension in the soil nails. The mobilization of tensile forces of soil nails in the active zone is mainly attributed to the reaction provided by the nail heads and facing, together with the friction between soil nails and the surrounding ground mass. The resistance against pullout failure of the soil nails in the passive zone is provided by the friction between soil nails and the surrounding ground mass. Ultimate pullout resistance is one of the key parameters for design of soil nailed structures.

In this chapter, the key factors governing pullout resistance are identified and reviewed, which include the effective normal stress or radial stress acting on the soil nail surface and the apparent coefficient of friction between nail and the surrounding ground. The former is further influenced by various factors, such as restrained soil dilatancy, arching effect, degree of saturation, and the nail installation method, whereas the latter is affected by the soil properties and soil nail characteristics. Given the complication of the soil nail pullout mechanism and the uncertainties involved, different

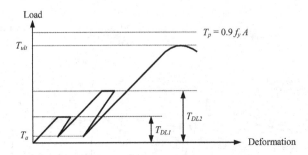

Schematic Diagram of Load-deformation Cycles of a Pullout Test

Testing Procedures

1.   The test soil nail shall be loaded in stages: from the initial load (T_a) via two intermediate test loads (T_{DL1} and T_{DL2}) to the maximum test load.

2. T_{DL1} shall be the allowable pullout resistance provided by the bond length of the cement grout sleeve of the test soil nail.

3. T_{DL2} shall be T_{DL1} times the factor of safety against pullout failure at soil-grout interface (F_{SG}).

4.   The maximum test load shall be 90 % of the yield load of the test soil-nail reinforcement (T_p) unless the ultimate ground-grout bond load (T_{ult}) is reached during the test.   Reinforcement size larger than that of the working soil nail should be used in the pullout test, where necessary, to allow the development of T_{ult} prior to reaching T_p.

5. T_a shall be T_{DL1} or 5 % of T_p, whichever is smaller.

6.   During the first two loading cycles, T_{DL1} and T_{DL2} shall be maintained for 60 minutes for deformation measurement.   The measurement at each of the cycles shall be taken at time intervals of 1, 3, 6, 10, 20, 30, 40, 50 and 60 minutes.   If the test soil nail can sustain the test load subject to the acceptance criteria given below, the load shall be reduced to T_a and the residual deformation shall be recorded, after which the test shall proceed to the next loading cycle.

7.   In the last loading cycle, the test load shall be increased gradually from T_a straight to the maximum test load and then maintained for deformation measurement.   The measurement shall be taken at time intervals of 1, 3, 6, 10, 20, 30, 40, 50 and 60 minutes.   If the test soil nail can sustain the test load subject to the acceptance criteria given below, the load shall be reduced to T_a and the residual deformation shall be recorded, after which the test is completed.

8.   If the test soil nail fails to sustain T_{DL1}, T_{DL2}, or the maximum test load in any cycle, the test shall be terminated and the soil nail movement against residual load with time shall be recorded.   The measurements shall be taken at time intervals of 1, 3, 6, 10 and, every 10 minutes thereafter over a period for at least two hours.   The measurements shall be taken for a longer period where considered necessary.

Acceptance Criteria

The test soil nail is considered to be able to sustain the test load if the difference of soil nail movements at 6 minutes and 60 minutes does not exceed 2 mm or 0.1 % of the bond length of the test soil nail.

Figure 6.5 Typical procedures and acceptance criteria for a pullout test in Hong Kong (GEO, 2008).

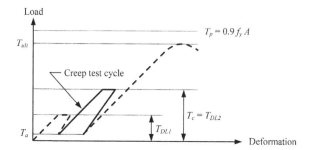

Schematic Diagram of Load-deformation Cycle of a Creep Test as part of a Pullout Test

Testing Procedures

1.   The procedures for a creep test are similar to those for a pullout test except that only one loading cycle is required.   Hence, it may be carried out as part of a pullout test.   Typical procedures for a pullout test and the definition of T_a, T_{DL1} and T_{DL2} are given in Figure 6.3.

2.   The test soil nail shall be loaded from T_a to the creep test load (T_c).

3.   The creep test load (T_c) is defined as the allowable pullout resistance provided by the bond length of the cement grout sleeve of the test soil nail times the factor of safety against pullout failure at soil-grout interface (F_{SG}), which is corresponding to the intermediate test load T_{DL2} for a pullout test.

4.   The creep period shall be deemed to begin when T_c is applied.   The load shall be maintained for 60 minutes for deformation measurement.   During the creep period, the measurement shall be taken at time intervals of 1, 3, 6, 10, 20, 30, 40, 50 and 60 minutes.

Acceptance Criteria

A test soil nail shall be considered acceptable when:

(a)   the difference of soil nail movements at 6 minutes and 60 minutes during the creep period does not exceed 2 mm or 0.1 % of the bond length of the test soil nail, and

(b)   the overall trend of creep rate (i.e., soil nail movement/log time) is decreasing throughout the creep period.

Figure 6.6 Typical procedures and acceptance criteria for a creep test in Hong Kong (GEO, 2008).

empirical, field testing, and analytical approaches have been developed and proposed for estimating the ultimate pullout resistance. Nonetheless, there is no unified method as reflected by the different approaches and codes of practice adopted by different places in the world. At present, the most common approach to estimate the pullout resistance of soil nails is still based on local experience, which includes published bond strength values, established correlations with soil properties, and analytical methods, such as the effective stress method. In practice, the design pullout value is to be verified by field pullout tests.

Various types of field pullout tests and their corresponding merits and limitations are also reviewed and discussed in this chapter. These include investigation, verification, acceptance, and creep pullout tests. Depending on the project requirements and the ground conditions, designers may select the most appropriate types of tests to suit the testing objectives.

REFERENCES

Bruce, D.A. & Jewell, R.A. (1987). Soil nailing: application and practice: Part 2. *Ground Engineering*, vol.20, no.1, pp. 21–33.

BSI (2010). Execution of Special Geotechnical Works: Soil Nailing (BS EN 14490: 2010). British Standards Institution, London, UK, 68 p.

BSI (2013). Code of Practice for Strengthened/Reinforced Soils: Part 2: Soil Nail Design (BS 8006–2: 2011). British Standards Institution, London, UK, 104 p.

Byrne, R.J., Cotton, D., Porterfield, J., Wolschlag, C. & Ueblacker, G. (1998). Manual for Design and Construction Monitoring of Soil Nail Walls. Federal Highway Administration, US Department of Transportation, Washington, D.C., USA, Report No. FHWA-SA-96-069R, 530 p.

Cartier, G. & Gigan, J. P. (1983). Experiments and observations on soil nailed structures. In Proceedings 8th European Conference on Soil Mechanics and Foundation Engineering, Helsinki, pp. 473–476.

Cheang, W.W.L. (2007). Static Behaviour of Soil Nails in Residual Soil. PhD thesis, National University of Singapore, Singapore, 364 p.

Cheung, R.W.M. & Shum, K.W. (2012). Review of the Approach for Estimation of Pullout Resistance of Soil Nails. Geotechnical Engineering Office, Civil Engineering and Development Department, HKSAR Government, Hong Kong, GEO Report No. 264, 49 p.

Chu, L.M. and Yin, J.H. (2005). Comparison of interface shear strength of soil nails measured by both direct shear box tests and pullout tests. *Journal of Geotechnical and Geoenvironmental Engineering*, vol. 131, no. 9, pp. 1097–1107.

Clouterre (1991). French National Research Project Clouterre: Recommendations Clouterre (English Translation 1993). Federal Highway Administration, US Department of Transportation, Washington, D.C., USA, Report No. FHWA-SA-93-026, p. 316.

Elias, V. & Juran, I. (1991). Soil Nailing for Stabilization of Highway Slopes and Excavations, United States Federal Highway Administration, Publication No. FHWA-RD-89-198, 210 p.

Franzen, G. (1998). Soil Nailing: A Laboratory and Field Study of Pullout Capacity. PhD thesis, Department of Geotechnical Engineering, Chalmers University of Technology, Göteborg, Sweden, 182 p.

GEO (Geotechnical Engineering Office) (2008). *Guide to Soil Nail Design and Construction (Geoguide 7)*. Geotechnical Engineering Office, Civil Engineering and Development Department, HKSAR Government, Hong Kong, 97 p.

Guilloux, A., Schlosser, F. & Long, N.T. (1979). Laboratory study of friction between soil and reinforcements. In Proceedings of the International Conference on Soil and Reinforcement, Paris, France, vol. 1, pp. 35–40.

Heymann, G. (1993). Soil Nailing Systems as Lateral Support for Surface Excavations. Master thesis, Faculty of Engineering, University of Pretoria, South Africa.

Heymann, G., Rohde, A.W., Schwartz, K. & Friedlaender, E. (1992). Soil nail pull out resistance in residual soils. In Proceedings of International Symposium on Earth Reinforcement Practice, Fukuoka, Japan, pp. 487–492.

Jewell, R.A. (1990). Review of theoretical models for soil nailing. In Proceedings of International Reinforced Soil Conference, Glasgow, Scotland, 10–12 September, pp. 265–275.

JHPC (Japan Highway Public Corporation) (1987). *Guide for Design and Construction on Reinforced Slope with Steel Bars.* Japan Highway Public Corporation, Japan, 33 p.

JHPC Japan Highway Public Corporation (1998). *Guide for Design and Construction on Soil Nailed Cut Slopes* (In Japanese). Japan Highway Public Corporation, Japan, 111 p.

Johnson, P.E. & Card, G.B. (1998). The Use of Soil Nails for the Construction and Repair of Retaining Walls. Transport Research Laboratory, TRL Report 373, 43 p.

Koivumaki, O. (1983). Friction between sand and metal. In Proceedings of the European Conference on Soil Mechanics and Foundation Engineering, Helsinki, vol. 2, pp. 517–520.

Lazarte, C.A., Robinson, H., Gomez, J.E., Baxter, A., Cadden, A. & Berg, R. (2015). Geotechnical Engineering Circular No. 7: Soil Nail Walls: Reference Manual. Federal Highway Administration, US Department of Transportation, Washington, D.C., USA, Report No. FHWA-NHI-14-007, 385 p.

Luo, S.Q., Tan, S.A. & Yong, K.Y. (2000). Pullout resistance mechanism of a soil nail reinforcement in dilative soils. *Soils and Foundations*, vol. 40, no. 1, pp. 47–56.

Mecsi, J. (1997). The load bearing capacity and the load-elongation diagram of soil anchors. In Proceeding of 8th European Conference on Soil Mechanics and Foundation Engineering, Hamburg, pp. 1327–1330.

Palmeira, M. & Milligan, G.W.E. (1989). Large scale direct shear tests on reinforced soil. *Soils and Foundations*, vol. 29, no. 1, pp. 18–30.

Pedley, M.J. (1990). The Performance of Soil Reinforcement in Bending and Shear. PhD thesis, University of Oxford, UK, pp. 200.

Phear, A., Dew, C., Ozsoy, B., Wharmby, N.J., Judge, J. & Barley, A.D. (2005). Soil Nailing: Best Practice Guidance. Construction Industry Research & Information Association, London, UK, CIRIA Report No. C637, 286 p.

Plumelle, C., Schlosser, F., Delage, P. & Knochenmus, G. (1990). *French National Research Project on Soil Nailing: Clouterre.* ASCE, Special Publication No. 25, ASCE, Reston, VA, pp. 660–675.

Potyondy, J. (1961). Skin friction between various soils and construction materials. *Géotechnique*, vol. 11, no. 4, pp. 339–353.

Pradhan, B. (2003). Study of Pullout Behaviour of Soil Nails in Completely Decomposed Granite Fill. M.Phil thesis, The University of Hong Kong, Hong Kong.

Pradhan, B., Tham, L., Yue, Z., Junaideen, S. & Lee, C. (2006). Soil nail pullout interaction in loose fill materials. *International Journal of Geomechanics*, vol. 6, p. 238.

Robinsky, E.I. & Morrison, C.F. (1964). Sand displacement and compaction around model friction piles. *Canadian Geotechnical Journal*, vol. 1, no. 2, pp. 81–93.

Sabatini, P.J., Pass, D.G. & Bachus, R.C. (1999). Ground Anchors and Anchored Systems (Geotechnical Engineering Circular No. 4). Federal Highway Administration, Washington, DC., Report No. FHWA-IF-99-015, 281 p.

Schlosser, F. (1982). Behaviour and design of soil nailing. In Proceedings of Symposium on Recent Developments in Ground Improvements, Bangkok, 29 Nov–3 Dec, pp. 399–413.

Schlosser, F. & Elias, V. (1978). Friction in reinforced earth. In Symposium on Earth Reinforcement, ASCE, Pittsburgh, pp. 735–763.

Schlosser, F. & Guilloux, A. (1981). Le forttement dans les sols. Revue Francaise de Gèotechnique, vol. 16, pp. 65–77 (in French).

Schlosser, F. & Juran, I. (1979). Design parameters for artificially improved soils. In Proceedings European Conference on Soil Mechanics and Foundation Engineering, Brighton, pp. 197–225.

Schlosser, F., Jacobsen, H.M. & Juran, I. (1983). General Report: Soil Reinforcement. Specialty Session 5. In Proceedings of the 8th European Conference on Soil Mechanics and Foundation Engineering, Helsinki, vol. 3, pp. 1159–1180.

Su, L.J., Chan, C.F., Shiu, Y. K., Cheung, T. & Yin, J.H. (2007). Influence of degree of saturation on soil nail pull-out resistance in compacted completely decomposed granite fill. *Canadian Geotechnical Journal*, vol. 44, no. 11, pp. 1314–1428.

Su, L.J., Chan, C.F., Yin, J.H., Shiu, Y.K. & Chiu, S.L. (2008). Influence of over-burden pressure on soil nail pullout resistance in a compacted fill. *Journal of Geotechnical and Geoenvironmental Engineering*, vol. 134, no. 9, pp. 1339–1347.

Tei, K. (1993). A Study of Soil Nailing in Sand. PhD thesis, University of Oxford.

Terzaghi, K. (1943). *Theoretical Soil Mechanics*. Wiley, New York.

Watkins, A.T. & Powell, G.E. (1992). Soil nailing to existing slopes as landslip preventive works. *Hong Kong Engineer*, March issue, pp. 20–27.

Wernick, E. (1978). Skin friction of cylindrical anchors in non-cohesive soils. In Symposium on Soil Reinforcing and Stabilising Techniques in Engineering Practice, Sydney, Australia, pp. 201–219.

Winterkorn, H.F. & Pamukcu, S. (1991). Soil stabilization and grouting. In *Foundation Engineering Handbook* (edited by Fang, H. & Van Nostrand Reinhold), Kluwer Academic Publishers, New York.

Xanthakos, P.P. (1991). *Ground Anchors and Anchored Structures*. Wiley.

Yin, J.H., Su, L.J., Cheung, R.W.M., Shiu, Y.K. & Tang, C. (2009). The influence of grouting pressure on the pullout resistance of soil nails in completely decomposed granite fill. *Géotechnique*, vol. 59, no. 2, pp. 103–113.

Chapter 7

Construction

7.1 INTRODUCTION

The major construction activities involved in a soil nailing project include excavation, nail installation, grouting, nail head and facing construction, surface and subsurface drainage provisions, etc. Successful construction of a soil nailed structure is attributed to many factors, including good project planning, good quality of materials, effective construction method, good workmanship, and construction safety. It is important that sufficient control measures, such as adequate supervision, risk mitigation measures, testing and monitoring commensurate with the scale and complexity of the particular project, are in place during the construction stage to ensure good quality of works and completion on time.

Proper supervision and control should be provided during all stages of soil nailing works, particularly those items of works that are difficult to be verified afterward, e.g. actual length of soil nail, quality of cement grout, integrity of couplers and corrosion protection measures, etc. Experienced site supervisory staff should therefore be engaged and provided with sufficient information and briefing for their appreciation of the geotechnical content of the works, safety requirements, key design assumptions, and the range of potential anomalies that could be encountered. A designer should review the validity of the assumptions critical to the design throughout the construction phase. Verification and review of the design should continue during construction stage when further information on the actual ground and groundwater conditions becomes available. This may necessitate refinements of the original design.

In this chapter, the key aspects concerning the construction of soil nailed structures are identified and discussed. These include various techniques for construction of soil nails and the associated components. The range of control measures and available engineering tools, including site supervision and various destructive or non-destructive tests, to enhance the quality of the works are presented. Construction difficulties that may be encountered in soil nailing works, together with the possible preventive or mitigation measures, are also highlighted. It is important that designers should carefully consider what may go wrong on a regular basis and develop an appropriate contingency plan during the course of construction works.

7.2 SAFE SYSTEM OF SOIL NAILING WORKS

Safety considerations must be given top priority in both the design and construction stages. All parties, including clients, designers, contractors, and frontline workers, should follow their local health and safety regulations during the construction works. An effective approach to avoid introducing a hazard to the workplace of soil nailing works is to establish a safe system of works. The system should be established as early as possible, such as at the planning and design stages, rather than when the hazards become real risks during the construction stage. The safe system of works is a process that involves participation by the clients, designers, contractors, and workers. Under the system, the project team is required to prepare a project risk management plan at the early stage of the project, so as to identify all the potential risk elements that a contractor may encounter, together with the respective feasible preventive or mitigation measures. The potential risk is then evaluated and a framework comprising design, specification, and operation activities is developed, which can either be used to prevent such hazards from materializing or be employed to mitigate their effects. Establishment of a safe system of works is the most effective risk control measure that can eliminate some of the hazards at source. The roles of the clients, designers, and contractors under the system are briefly described in the following.

Client: A client is the party who finances the project and is able to influence many major decisions involved in the project. In this regard, the client plays a critical role in stipulating safety-related requirements. A client should manage their project risks in a responsible and considerate manner. A good client should be clear about what the high-level risks are during construction and which key risks during the operation and maintenance phase would need to be managed after completion.

Designers: Designers should identify all the potential construction hazards and provide all the key information that the contractors need to be made aware of. They are in a position to make major contributions to safety and health by hazard identification and elimination, and by risk reduction during the design stage. They should work closely with other duty holders to come up with a design that duly considers buildability and safety aspects. Engineering drawings, design reports, or models can be used to present their design ideas with other duty holders. Designers should deal with the relevant safety and health issues for the project.

Contractors: Contractors are responsible for carrying out the works. They play a critical role in ensuring that hazards identified both prior to and during the works are properly addressed. They are responsible for the planning, management, and coordination of construction works. They should be able to effectively manage safety and health issues during the construction stage. They should ensure that they receive all the essential information pertaining to the identified risks as early as possible. It is preferable for contractors to receive all the relevant information concerning the identified risks at the earliest time, so that they can have sufficient time to review the information and prepare an appropriate method statement and undertake a risk assessment.

For more details about construction safety and health issues, readers may refer to the relevant guidelines being promulgated in various places, such as Construction Design Management and the Design for Safety currently promulgated in the UK and Hong Kong, respectively (e.g. HKSAR, 2013a, b).

7.3 SOIL NAIL INSTALLATION TECHNIQUES

7.3.1 General

There are two principal techniques of soil nail installation, namely (i) "drill-and-grout" and (ii) direct driving methods. The former is the most common nail installation method, where upon completion of hole drilling and insertion of nail reinforcement, the drillhole is filled with cement grout either by gravity or under pressure. A variation of the "drill-and-grout" method is the self-drilling technique in which the nail reinforcement, in the form of a hollow bar, is used as a self-drilling rod and grout is used as a flushing medium such that the drilling and grouting process can be carried out simultaneously. In the latter direct driving method, the soil nail may be installed by percussive, vibratory, or ballistic means. The selection of the suitable installation method depends on a number of factors, including ground conditions, prior experience of the contractor, availability of the construction plant, required construction rate, site constraints, cost, safety and environmental considerations, etc.

7.3.2 Drilling methods

Most of the soil nail installation involves the use of drilling rigs. If the working space is adequate, crawler type mini-pile pneumatic percussive drilling rigs powered by air compressors may be used for drilling (Figure 7.1). Large hydraulic-powered rigs may also be used to drill holes for long soil nails, say, up to 30 m. These rigs can drill the entire length of hole in a single pass without having to add sectional augers. Nonetheless, this type of drilling requires sufficient space to accommodate the large-scale machinery. For steep cutting or at locations where working space is limited, a modified light-weight rotary or pneumatic percussive drilling rig with the motor mounted onto a fixed leader, powered by an air compressor, is preferred (Figure 7.2). Air is commonly used for the flushing medium during the drilling operation. This drilling technique is very common in Hong Kong. The

Figure 7.1 Crawler type mini-pile drilling rig.

Figure 7.2 Light-weight rotary drilling rig.

Figure 7.3 Special small-scale drilling rig.

rigs are usually operated by one or two operators and the entire set-up can be accommodated on benches of only a few meters wide.

Under special circumstances where the working space and access are extremely limited, an even shorter and tailor-made drilling machine as shown in Figure 7.3 could be used for drilling. This special drilling rig has a leader less than 1 m long. However, additional manpower may be required to connect and fix the drill rods of 500 mm length during the drilling operations. A hand-held coring machine is also a feasible option for drilling works with limited space (Figure 7.4). Coring machines generate less noise, dust, and vibration than their percussive counterparts, but the speed of coring is very slow, and the maximum length of coring is less than about 10 m.

In Germany and the United States, auger drilling is the most common technique used to construct an open hole. In Hong Kong, pneumatic percussive drilling with air as the flushing medium, rotary, and hand-held rotary drilling are mostly used for hole formation, and temporary casing is normally not required in typical *in situ* soils. However, if cavities are encountered in loose soils during the drilling operation, which could be detected by a sudden drop in the pressure of the air compressor, eccentric overburden drilling technique, such as ODEX (Figure 7.5), may be considered. This technique enables a temporary casing to be advanced into a hole concurrent with drilling in order to prevent the hole from collapsing in difficult ground conditions. The reamer bit swings out at the bottom of the temporary casing and drills a hole slightly larger than the outside diameter of the casing, hence facilitating its advancement. The casing helps to support the ground and the return of the flushing air. However, some cuttings will be flushed up between the casing and the ground as the drillhole is slightly larger than the casing. This could cause some disturbance to the surrounding ground as well as ground loss.

Figure 7.4 Hand-held drilling machine.

Figure 7.5 Drill bit for eccentric overburden drilling.

Figure 7.6 Ring bit for concentric drilling.

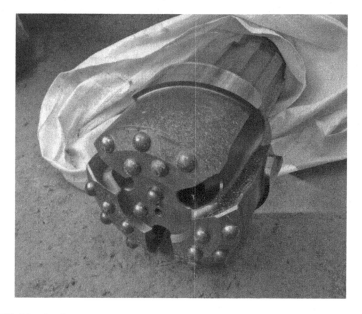

Figure 7.7 Pilot bit for concentric drilling.

Recently, a concentric drilling technique, as shown in Figures 7.6 and 7.7, has been developed which does not have any eccentric parts and works with a pilot bit system together with a ring bit. The ring bit fully covers the pilot bit and guides the air to return inside the casing, hence the disturbance to the surrounding ground would be minimized. Without excessive torque exerted on other parts of the drilling rig as compared with the eccentric overburden drilling technique, the concentric drilling technique can work well in difficult ground conditions, such as very loose ground or bouldery ground. The main drawback of this technique is that the ring bit together with the casing shoe is welded to the permanent casing, which has to be left underground when the pilot bit and down-the-hole hammer are retrieved. Moreover, a longer leader of about 2.5 m length has to be used to accommodate the pilot bit system and a higher air pressure, up to about 17 bar (1,700 kPa), is required.

7.3.3 Drilling followed by grouting of soil nails

The "drill-and-grout" nail installation technique is the most common method adopted for soil nail installation. This involves constructing a drillhole at a predetermined location, inclination, orientation, and length, followed by placing the nail reinforcement and the grouting operations. The drillhole may be unsupported if the soil is competent or else a temporary or permanent casing would be required. The open drillhole, typically with diameter ranging from 100 mm to 200 mm, is commonly formed by drilling using one of the available methods such as auger, rotary, percussion, and rotary-percussion as described in Section 7.3.2. In competent ground materials, auger, rotary, or rotary-percussion with flushing medium would normally be employed, while for weak materials such as loose fill, auger or rotary methods with casing could be used.

Upon the completion of drilling works, the soil nail reinforcement, with centralizers provided at regular intervals along its length, is inserted into the bottom of the drillhole. Grouting is then carried out by gravity or under a low pressure (normally 5 to 10 bar, i.e. 500 to 1,000 kPa) from the bottom of the drillhole, through a grout pipe for the open-hole installation method or through grout pipe or hollow stem for the cased installation method. Grouting should continue until fresh cement grout emerges from the top of the drillhole with no signs of contamination or inclusion. This would ensure good integrity of the cement grout. The most common grouting method adopted in the United States and Hong Kong is by gravity grouting. However, if poor ground condition is encountered or a higher bond strength is required, grouting under a suitable pressure may be considered.

7.3.4 Simultaneous drilling and grouting of soil nails

Drilling and grouting simultaneously, also referred to as self-drilling soil nail, is another form of "drill-and-grout" nail installation technique. In this method, the nail reinforcement, which also acts as the drilling rod, is equipped with a sacrificial drill bit. The selection of drill bit depends very much on the ground conditions. The nail reinforcement is usually a hollow bar to facilitate the injection of cement grout into the drill bit. Throughout the installation process, the grout is injected through the hollow bar concurrently with the drilling. The cement grout serves as the flushing medium and its fluid pressure will maintain the stability of the drillhole. Rotary percussive drilling is usually adopted for the method. As the diameter of the self-drilled hole is usually smaller than that formed by conventional drilling techniques, hand-held equipment is commonly employed for this method. This installation technique is usually used for temporary applications.

7.3.5 Driven soil nails

This type of nail installation technique is commonly employed in France and Germany. In this technique, the soil nails are driven directly into the ground by means of percussive, vibratory, or rotary methods. As the nail reinforcement is driven directly into the ground, the reinforcement would be in direct contact with the ground and the soil surrounding it would be displaced during the installation process. The nail reinforcement should have sufficient strength, including bending stiffness, to withstand the impact and buckling force during installation. In general, this type of soil nail has a relatively low pullout resistance. However, if a higher resistance is needed, post-installation grouting may be carried out. Special nail reinforcement with an axial channel can be used to allow for grouting of the nails. According to experience in the United States, the technique is economical, and the installation is rapid (four to six nails per hour). The maximum length of installation can be up to 20 m in suitable ground conditions. Its application may, however, be limited by the ground conditions, e.g. it is not suitable for ground with obstructions such as boulders. As this type of installation cannot provide good corrosion protection to the nail reinforcement, driven nails are often used for temporary applications.

7.3.6 Ballistic or launched soil nails

The technique is a variation of the driven nail method which was originally developed by the British military. It was subsequently declassified for civilian applications, including soil nail installation. In this technique, ballistic or launched soil nails of 6 m length are fired into the ground in a single shot at very high speed by means of high-pressure compressed air. The nail reinforcement can be in the form of a solid bar, a threaded bar, or a pipe. Steel bars can be left as mill finished, galvanized, epoxy coated, with a plastic sheathing, or a combination of various coatings. As the nail reinforcement enters the ground at a very high speed (about 400 km/hour), a shock wave is developed in front of the nail tip resulting in soil moving away from the main shaft. The nail reinforcement can therefore be installed into the ground without significant abrasion or coating damage. The ballistic equipment can be mounted on a range of vehicles, tracked excavators, or slung from a crane. The ballistic nails were first adopted in the UK in 1989. Subsequently, track records of such installation in other places, such as the United States, Canada, Australia, New Zealand, and some European countries, are found. Nonetheless, the application of ballistic nails is still not common as there are only a few ballistic nail launchers available in the world. Moreover, due to the difficulty in controlling the length of nail that can penetrate into the ground, ballistic nails are usually used for temporary applications, such as landslide repair works.

7.4 CONSTRUCTION SEQUENCE

7.4.1 General

As the majority of soil nails are installed using the "drill-and-grout" method, the construction sequence is discussed in detail in this section. In site formation works, soil nailing is normally carried out in association with excavation involving the following steps:

(a) excavation (top-down),
(b) drilling of holes,
(c) insertion of nail reinforcement attached with accessories, such as centralizers,
(d) grouting, either by gravity or under pressure,
(e) construction of temporary or permanent nail head/facing, and
(f) construction of subsequent level by repeating steps (a) to (e).

If soil nails are installed into existing geotechnical structures such as slopes, retaining walls, or embankments, the following sequence of works will be involved:

(a) drilling for holes (top-down or bottom-up),
(b) insertion of nail reinforcement attached with accessories, such as centralizers,
(c) grouting, either by gravity or under pressure, and
(d) construction of permanent nail head/facing.
(e) construction of subsequent level by repeating steps (a) to (d).

The critical construction activities are further elaborated in the following sections.

7.4.2 Excavation

Prior to commencement of the excavation works, a monitoring system (e.g. for monitoring ground deformation, groundwater, deformation of nearby structures and utilities, etc.), as well as temporary surface and subsurface drainage should be in place for construction control and monitoring the impact of the works on the surrounding area and preventing surface or subsurface water from adversely affecting the works. Moreover, a system to control the tolerance of the excavation, such as the height and gradient of excavation, should also be established. Excavation usually involves a bulk operation followed by trimming of the cut face. Unless the overall height of excavation is small, the works are usually carried out in stages. The height of the exposed excavated face, usually 1 m to 2 m, is determined with due consideration given to temporary stability. An excavated face of good quality is essential for construction of the slope facing. Any loose materials and localized protrusions should be removed or trimmed from the excavated face and backfilled with suitable materials. This facilitates good contact of the facing with the excavated face. After installation of a row of soil nails, the subsequent excavation should progress only when the temporary stability of the excavation is adequate. Soil nail heads and slope facing should be constructed before the next stage of excavation unless the temporary stability of the soil nailed excavation in the absence of soil nail heads is deemed to be adequate. The sequence and timing of installing soil nails, constructing soil nail heads and facing, and subsequent excavation should be monitored and controlled to fulfil the above requirements. In order to minimize the risk of face collapse, it is common practice to shorten the period between excavation, soil nail installation, and head/facing construction as far as possible.

During the staged excavation, a good opportunity is provided for the designer to review the actual ground profile and groundwater conditions from the exposed face. This new information should be reviewed vis-à-vis the design assumptions made on the basis of pre-design site investigation. If necessary, soil samples may be retrieved for further testing and suitable design amendments be made accordingly.

7.4.3 Drilling

As discussed in Section 7.3, an open hole can be formed in the ground by auger, rotary, percussion, or rotary-percussion drilling. If the soil has sufficient shear strength, air would normally be used as the flushing medium for rotary, percussion, and rotary-percussion drilling techniques. Under special circumstances, water or cement grout is used as the flushing medium, due to its greater density, to maintain the stability of the drillhole. The drillhole should be formed to a sufficient length to accommodate the soil nail reinforcement. It is important to ensure that the alignment of the drillholes is within the permitted tolerances in order to avoid possible clashing of soil nails, in particular for closely spaced or long soil nails, or soil nails with different inclinations and bearings. It is common practice to check the inclination and bearing of drillholes with the use of a protractor and compass

Figure 7.8 Setting the inclination of the drilling rig using a protractor.

on the drill rods (Figure 7.8). It is of paramount importance to control and check the initial inclination and bearing of the drillhole. If accurate measurements of the inclination and bearing of the drillhole along its length are needed, special equipment, such as Eastman camera, may be employed.

For long soil nails, the drilling rate and compressed air pressure should be suitably controlled to minimize the eccentricity that can be produced by the inclination of the drill rods to dip or droop downward. Otherwise, this may cause misalignment of the drillhole or unduly enlarge the diameter of the drillhole and cause hole collapse. Drillholes in soil should be kept open only for a short period of time. The longer the hole is left open, the greater the risk of collapse. Under special circumstances, such as drilling through an unstable soil stratum, a temporary steel casing may be used to maintain hole stability and enhance buildability. Grouting would be carried out prior to retrieving the casing.

Drilling under water should be treated with caution because the hole is more susceptible to collapse than that in dry condition. It may also cause disturbance to the adjacent ground, which can weaken the bond strength of the soil nail. In difficult ground conditions, suitable dewatering measures may need to be considered to facilitate the drilling of soil nails. The effects of any dewatering, such as ground settlement and differential settlement, should be duly assessed and suitable mitigation measures should be implemented as appropriate.

Sometimes, soil nails are used to stabilize existing geotechnical structures, such as reinforced concrete retaining walls. In this case, before commencing the drilling works, safety precautions should be taken to avoid damaging the steel bars in the reinforced concrete wall, e.g. by using a metal detector to determine the locations of steel bars.

Tools such as Drilling Process Monitor (DPM) (Yue et al., 2004) may be used to record the information on the hole drilling process. The DPM system collects and processes real-time drilling data for monitoring the works. The system could be used to deduce the likely ground conditions at the soil nail locations provided that sufficient calibration of the DPM data with known ground conditions is undertaken.

7.4.4 Insertion of soil nail reinforcement

The soil nail reinforcement should be handled with care to avoid any damage (Figure 7.9). The integrity of centralizers and corrosion protection measures, such as hot-dip galvanizing, corrugated plastic sheathing, and heat-shrinkable sleeves, should be checked prior to insertion into the drillhole. Pitting spots should not be tolerated as these may lead to severe corrosion under aggressive ground conditions. Where couplers are used to join sections of reinforcement, care should be taken to ensure that the couplers comply with the specification requirements.

Drillholes should be kept clean, otherwise the integrity of the cement grout annulus will be affected. Simple tools such as a mirror and high intensity light are generally good enough for inspecting the drillhole for cleanliness. For long drillholes, closed-circuit television (CCTV) may be used. If

Figure 7.9 Insertion of nail reinforcement.

obstructions are encountered during insertion of the soil nail reinforcement into the drillhole, the reinforcement should be withdrawn, and the obstruction removed before the reinforcement is re-inserted.

During insertion of the reinforcement, it should not be placed all the way to the bottom of the drillhole. A minimum clearance as specified by the designer should be maintained between the end of the reinforcement and the bottom of the drillhole so that there will be adequate cement grout cover to the soil nail reinforcement. At the same time, the reinforcement should also project a sufficient length beyond the face to allow proper connection with the nail head or facing system.

7.4.5 Grouting

Grouting should be carried out as soon as possible, for example, on the same day when the soil nail reinforcement is inserted into the drillholes, in order to minimize the potential for hole collapse. The grout usually comprises a neat cement grout with a water-cement ratio of 0.4 to 0.5. Grouting may be carried out either by gravity or under pressure. In addition, it can be carried out prior to reinforcement insertion or afterward, although the latter is more common. A grout tube is inserted to the bottom of the drillhole and grouting should continue until the cement grout emerging from the top of the hole becomes uncontaminated. This helps to ensure the integrity of the cement grout. The grout tube may be withdrawn during the grouting process or permanently left in place as part of the works. If the grout tube is to be withdrawn during the grouting operation, the withdrawal rate should be in such a manner that the end of the tube is below the prevailing grout surface during the entire process. For driven nails, a grout tube can be attached to the head of the reinforcement and the grout is injected through the tube upon completion of driving. Similarly, for self-drilled soil nails with hollow reinforcement, the grout may be injected by gravity or under pressure simultaneously with the drilling or subsequent to completion of drilling. The actual grouting pressure to be adopted is usually established through site trial.

For grouting of soil nails with corrugated plastic sheathing, excessive deflection and distortion of the sheathing between supports may occur if the inner annular space between the wall of the sheathing and soil nail reinforcement is grouted first, particularly where the centralizers are not strong enough to support the weight of the wet grout. However, grouting outside the sheathing first may result in floating of the sheathing. Appropriate measures, such as concurrent grouting inside and outside the sheathing using a Y-shaped connector to the grout pipe, should be taken.

Grouting under water should be avoided as far as practicable because the integrity of cement grout may be adversely affected. Provision of temporary dewatering measures, e.g. raking drains, should be incorporated in the design if it is expected that the grouting operation would be affected by groundwater. Where the drillholes encounter upon a persistent groundwater regime, dewatering measures should be implemented prior to the grouting operation. The effects of any dewatering on adjacent ground and facilities should be duly assessed and mitigation measures should be taken to alleviate the effects as appropriate.

7.4.6 Construction of soil nail heads and slope facing

Soil nail head typically comprises reinforced concrete pad, steel bearing plate, and hexagonal nut. The bearing plate and the nut, which may be embedded within or bedded onto the concrete pad, provide a connection between the soil nail reinforcement and the concrete pad. Soil nail heads can be recessed into the slope to minimize visual intrusion (Figure 7.10). Particular attention should be paid to ensuring the integrity of soil nail heads, especially at the location beneath the bearing plate where honeycombing of concrete can occur. It is a good practice to construct the soil nail heads in two stages if shotcrete (i.e. sprayed concrete) is used. The first stage concreting should be applied to a specified height above the intended base level of the steel bearing plate. The bearing plate should then be hammered into place and the nut tightened onto the soil nail reinforcement before commencement of the second stage concreting. For quality control, some soil nail heads should have the bearing plate removed in order to check the workmanship of the concreting.

Figure 7.10 Excavation for recessed soil nail head.

The slope facing generally serves to provide the slope with surface protection and minimize erosion and other adverse effects of surface water flow on the slope. The facing should be constructed soon after completion of excavation and nail installation in order to avoid local failure of the slope surface. A slope facing may be soft, flexible, hard, or a combination of the three. A soft facing is normally a vegetation cover in conjunction with a steel wire mesh and an erosion control mat, which is either biodegradable or non-biodegradable (Figures 7.11 and 7.12). The facing should be installed in accordance with the design requirements. Apart from minimizing erosion, the erosion control mat and steel wire mesh can facilitate the establishment of the vegetation cover. In general, the erosion control mat and steel wire mesh are fixed onto the slope surface by pattern pins. The selection of suitable vegetation species depends on a number of factors, such as the local climate, soil type and thickness, rainfall pattern, and slope orientation. Grass, groundcover, climbers, and shrubs are preferred to trees for planting on steep slopes (i.e. slope angle exceeding 35°) because of potential stability issues of tree planting.

A flexible facing, which is usually a proprietary product, contains coated metallic or geosynthetic netting. This should be installed in accordance with the manufacturer's recommendations. In general, a flexible facing is connected to the soil nail head plates and slope face with pattern pins. Particular attention should be paid to the installation details at the connection between the facing and the soil nail heads, especially at the upper row of soil nails where a secure connection is of paramount importance.

Figure 7.11 A soft slope facing with non-biodegradable erosion control mat.

Figure 7.12 A soft slope facing with biodegradable erosion control mat.

Upon completion of flexible facing installation, a vegetation cover could be provided to improve the aesthetic appearance and enhance the erosion resistance. However, if the facing is steep (i.e. slope angle exceeding 55°), other proprietary system or a crib facing may need to be used instead.

Hard facing, which includes sprayed concrete, reinforced concrete, and stone pitching, is particularly suitable for steep slopes and excavations (Figures 7.13 and 7.14). If steel mesh is used in conjunction with sprayed concrete, precautionary measures should be taken to avoid the creation of voids during concrete spraying. Sometimes, a thin layer of concrete may be sprayed on the exposed face immediately after the excavation for temporary protection.

7.5 CONSTRUCTION OF SURFACE, SUBSURFACE, AND FACE DRAINAGE

Surface water runoff and existing groundwater condition should be properly controlled to ensure satisfactory performance of a soil nailed structure, both during construction and throughout its design life. Suitable surface drainage provisions (e.g. crest channels, intercepting ditches, and stepped channels) and subsurface drainage provisions (e.g. raking drains and geocomposite drainage) should be provided to soil nailed structures based on the actual site conditions. The collected surface and subsurface water

Figure 7.13 Hard slope facing with sprayed concrete and planter holes.

Figure 7.14 Hard slope facing with masonry stones.

should be directed to a proper discharge outlet to prevent adverse effects of concentrated water flow on slope stability.

During construction, sufficient temporary drainage should be provided at all times, especially during the wet season, to avoid any adverse effects of uncontrolled concentrated water ingress or surface water flow. The temporary site drainage should be properly maintained and cleared of any blockage to ensure that the drains remain functional during heavy rainfall. The contractor should be encouraged, or required where appropriate, to construct part of the permanent drainage measures, e.g. crest drain and the associated discharge points, at an early stage of the works in order to enhance the temporary drainage provisions.

Subsurface drainage (raking drains) in the form of horizontal or gently inclined drain pipes may be installed to control the groundwater pressure behind the soil nailed structure. Drain pipes typically of 50 mm in diameter PVC slotted or perforated tubes with impermeable invert are installed upward at $5°$ to $10°$ to the horizontal. The pipes are normally longer than the length of soil nails such that the groundwater can be drained away before entering the reinforced soil mass. Drain pipes should be installed after nail installation to prevent potential intrusion of cement grout into the pipes. During the installation of drain pipes, due attention should be paid to avoid damage to the installed soil nails adjacent to the drains.

Face drainage in the form of vertical geocomposite strip drains is very common in the United States and some European countries. Geocomposite strip drains, of 200 mm to 300 mm in width, are installed behind the slope facing (e.g. sprayed concrete) to prevent the build-up of groundwater pressure. The strip drains are typically installed vertically from the top to the bottom, at specified intervals usually the same as the nail spacing, upon completion of each lift of excavation. The strip drains should be securely fixed on the excavated face with the geotextile side placed against the ground. For a highly irregular excavated face, the placing of strip drains could be difficult or even impractical. Where sprayed concrete facing is used, precautionary measure should be taken to prevent the drains from being contaminated by the shotcrete. Otherwise, its drainage capacity would be adversely affected. Strip drains are usually spliced at the bottom of each excavation lift with some overlapping to ensure that the water flow is not impeded. If the geotextile component of the drains becomes torn or punctured, the damaged section should be repaired or replaced. At the base of the soil nailed structure, strip drains are usually connected to footing drains below the finished grade or to weep holes that penetrate the structure. Weep hole locations should coincide with the drain locations. A filter fabric is usually placed against the weep holes to prevent clogging. Footing drains are comprised of perforated pipes embedded in drainage gravel surround. In Hong Kong, an alternative face drainage comprising weep holes is more common than the use of strip drainage. Weep holes comprise short pipes passing through the slope facing that can transmit water away from the ground behind. The spacing of weep holes depends on the groundwater condition, and is typically about 1 m to 3 m.

7.6 DUST AND NOISE CONTROL

Precautionary measures should be taken to ensure the soil nail installation would not generate excessive dust and noise to workers and general public in the vicinity of the site. The limits of dust and noise should be prescribed in accordance with the local regulations. In Hong Kong, dust suppression is conventionally carried out by spraying water and covering the surrounding areas of the drilling rig with tarpaulin sheets as shown in Figure 7.15. Recently, a new system comprising a portable air blower connected to a bucket of water through a duct has been developed through which the dust generated from the drilling rig is sucked (Figure 7.16).

7.7 TEMPORARY TRAFFIC ARRANGEMENT

If the construction works affect traffic flow, e.g. upgrading an existing roadside slope along a narrow carriageway as shown in Figure 7.17, a temporary traffic arrangement (TTA) scheme is normally required. TTA is usually

Figure 7.15 Dust control by tarpaulin sheet cover.

Figure 7.16 Dust control by conveying duct.

Figure 7.17 Narrow road on hilly terrain in Hong Kong.

undertaken for the construction works affecting existing roads with the following purposes:

- For loading and unloading activities.
- For working area.
- For storage of materials and plant.

Sometimes, temporary closure of traffic lanes may be required for construction activities. However, this should be carried out during non-peak hours. If a relatively long period of land closure is necessary, nighttime closure may be considered. The final developed TTA scheme should comply with the local requirements and be compatible with the site settings.

7.8 POTENTIAL CONSTRUCTION PROBLEMS

7.8.1 General

On occasions, problems are encountered during installation of soil nails. These may be associated with hole drilling, grout intake, integrity of concrete nail heads, and working space. Problems that could be encountered during drilling include slow progress or jamming of drill bits (especially in bouldery ground, differential weathering profile involving alternating zones of soil and rock, or highly fractured rock), drillhole becoming out of alignment (e.g. a downward drooped profile of the drillhole in very long soil nails, which may cause difficulty during the insertion of nail reinforcement as well as the subsequent grouting operation due to the increased likelihood of air being trapped and not fully displaced by the fluid grout from the bottom of the drillhole), hole collapse, undesirable ground movement, or excessive vibration caused to nearby sensitive receivers, etc. Permeable ground or presence of extensive underground voids may lead to excessive grout leakage. This could jeopardize the integrity of the grout annulus and hence the performance of a soil nailed structure. Lack of working space is also another common construction problem in well-developed and highly populated areas. During the design stage, it is important for a designer to thoroughly appreciate the potential construction problems that may be encountered so that the necessary preventive or mitigation measures could be devised and implemented.

7.8.2 Drilling of holes

A study of drillhole measurements in respect of alignment and diameter was conducted in Hong Kong (Lui, 2009). The results indicate that the measured and design values of hole inclination, bearing, and diameter are subject to different degrees of disparity. In particular, the disparity in hole inclination increases as the length of hole increases. It is probably not uncommon to have deviations of hole inclination from the design value exceeding $\pm$ 2° (which are beyond the stipulated requirement in Hong Kong). In practice, such specification requirement is found to be not always achievable, particularly for long nails or at sites with relatively less dense geological material. The disparity in hole bearing does not show a consistent trend either to the left or the right, and it may change course along the length of a hole. The study identifies some factors that may potentially affect the discrepancies and create large alignment deviations for drillholes. These include:

- Length of drillhole, since a significant length tends to be associated with greater alignment deviations.
- Presence of a considerable amount of relatively less dense geological material, i.e. colluvium, residual soil, or filling.
- Size of drill bit, since the use of a longer drill bit tends to give rise to greater deflections.

Soil nail drillholes in weak materials, loose soil, or bouldery ground are liable to collapse during formation of the open hole. Sites with high groundwater tables will exacerbate the potential hole collapse problem. Longer and larger diameter drillholes are comparatively more susceptible to collapse for given ground conditions and groundwater conditions. The experience and skill of the drillers, as well as the wear and tear condition of the drill bit and the applied compressed air pressure, can also be important factors in avoiding hole collapse.

Collapse of drillhole would hamper progress and could result in loosening or movement of the surrounding ground, which could adversely affect adjacent sensitive facilities and the pullout resistance of soil nails. Hole collapse during the grouting operation can also lead to integrity problems in the grout annulus, such as inclusions of soil and voids in the grout.

Temporary or sacrificial subsurface drains are sometimes used to deal with local heavy seepages that could promote hole collapse. This approach has given rise to mixed results in practice as the effectiveness of subsurface drains is very much dependent on the variability and complexity of the actual groundwater regime. Subsurface drains may potentially lead to drawdown of the groundwater level and possible ground movement. Movement monitoring should be carried out as considered necessary, especially where there are sensitive facilities in close proximity to the soil nailing site.

Other means to deal with poor ground and local high groundwater table include the repeated "drill-and-grout" method, in which a thicker grout (i.e. lower water to cement ratio) may be used. Drilling with a permanent casing using the eccentric or concentric drilling technique may be adopted to provide support to the unstable soil strata. The *Tube-a-Manchette* method has also been used to improve the ground condition by means of pressure grouting. It should be borne in mind that the above expediency may not necessarily be applicable in all ground conditions and that suitable site trials are needed to confirm their effectiveness on a case-by-case basis.

7.8.3 Grout loss

Presence of extensive underground voids or permeable ground may lead to excessive grout leakage during grouting operation. Excessive grout leakage implies difficulty in forming an intact cement grout annulus and may give rise to serious grout integrity problems (Figure 7.18). Excessive migration of cement grout into the soil may also increase the risk of groundwater damming and ground contamination. For excessive grout leakage to occur, the ground surrounding the soil nail must be highly permeable to fluid grout. Such ground is either generally permeable in which case the grout would set in the pore space within the soil surrounding the soil nail or may contain discrete pathways in which case the grout may migrate well away from the soil nail before setting. Based on field experience, some possible geological conditions that may contribute to excessive grout leakage are summarized in Table 7.1.

The conditions set out in Table 7.1 may serve as an initial screen to help assess the possibility of grout loss problems, although they may not necessarily mean that excessive grout leakage will occur at a particular site, and vice versa. In general, drainage lines with associated colluvium are potentially most vulnerable to grout loss during soil nail construction. Sites with open joints and/or erosion pipes are also comparatively vulnerable to grout loss.

Figure 7.18 View within a soil nail annulus devoid of grout.

Table 7.1 Geological conditions prone to excessive grout leakage

Geological Conditions	Potential Factors Leading to Excessive Grout Leak
Permeable Coarse Materials	• Ground containing a significant amount of boulders, cobbles, and gravel. • Colluvium and fluvial deposits containing significant amount of coarse materials and relatively low silt or clay content in the matrix.
Adverse Geological Setting	• Erosion pipes. • Boundaries within colluvium, and between colluvium and in situ material and within corestone-bearing saprolite. • Open joints, faults, and shear zones and other discontinuities (e.g. zones of hydrothermal alteration, etc.) that are weathered and eroded, and have an open soil structure. • Landslide scars, tension cracks, and other features following ground deformation as these may include voids within transported or in situ materials. • Drainage lines intersecting slopes within which colluvium may be present, erosion pipes may be developed, and preferential groundwater throughflow as indicated by seepage locations/horizons. • Ground with degenerated root system of dead trees. • Old buried streamcourse or drainage lines.

If excessive grout leakage occurs, it may not be possible to fill up the drillhole with grout. In such circumstances, it would not be advisable to continue injecting a large quantity of grout into the drillhole. Uncontrolled grout injection may adversely affect the subsurface water flow paths, lead to blockage of sub-surface drainage provisions in the nearby slopes, or undue influence on utilities. Excessive grout leakage, especially in the case of closely spaced soil nails in close proximity to subsurface drainage lines, is liable to cause damming effects and could lead to more adverse groundwater conditions behind the grouted zone and the adjacent areas.

A simple way to detect underground holes or voids is to monitor the drilling operation of soil nailing works. Unreasonably high penetration rate of drilling or lack of air return in the drilling process may indicate the possible presence of underground voids. With the aid of other investigation techniques, such as CCTV survey and geophysical survey (e.g. gravity or resistivity survey, etc.), one can determine the nature, extent, and location of the voids. It is good practice to monitor the amount of cement grout take by recording the volume of cement grout placed in each drillhole. In case of excessive grout leakage, the designer should be notified promptly in order to determine the appropriate action to be taken. Possible mitigation measures to avoid or minimize excessive grout leakage include:

- Use grout of a lower water to cement ratio (e.g. 0.35 to 0.4).
- Adopt staged grouting, i.e. grouting is carried out intermittently to allow time for grout set.

- Repeated drilling and grouting, i.e. the drillhole is grouted up and then re-drilled.
- Carry out pre-grouting in the vicinity of the drillhole.
- Use chemical grout or cement grout with an expanding admixture.
- Provide a sleeve over the location of grout leakage using a PVC or permanent steel casing.

In parallel with the above measures, one should also monitor the quantity of grout injected into the drillholes and assess the effects of any excessive grout leakage into the ground on the hydrogeology of the nearby area.

7.8.4 Honeycombing in concrete soil nail heads

A soil nail head typically comprises a reinforced concrete pad, steel bearing plate, and nuts. Its primary function is to provide a reaction for individual soil nails to mobilize tensile force. It also contributes to local stability of the ground near the slope surface between soil nails. Soil nail heads are sometimes constructed by means of ready-mix concrete (e.g. this was the practice in Hong Kong in the early 1990s). However, there are more restrictions in the use of ready mixed concrete as compared to shotcreting. Normally, the volume of concrete required for the construction of soil nail heads is relatively small and does not facilitate the optimal and timely use of the concrete trucks' delivery. The concreting operation would become even more difficult for slopes with no vehicular access. Under such circumstances, manual transportation, i.e. bucket by bucket, would be the only means to cast the nail heads, in which case it would be very difficult to control the quality and workmanship of the works. Soil nail heads are commonly constructed by shotcreting nowadays. Dry mix is adopted in shotcreting since it is readily available in all concrete plants. Quality assurance of soil nail head shotcreting is further enhanced by stipulating the qualification and experience requirements of the nozzleman. Quality control testing of the compressive strength of concrete cores recovered from a test panel constructed at the same time as shotcreting is also specified. An advantage of the dry-mix process is that it can be applied to sites with no vehicular access as the operation relies on the pressure provided by an air compressor placed at a convenient location away from the works. Particular attention should be paid to ensuring the integrity of soil nail heads, especially at the location beneath the bearing plate where honeycombing can be formed easily. For soil nail heads to be constructed using shotcrete, it is good practice to construct the soil nail head in two stages. In the first stage, shotcrete should be applied to a specified thickness above the intended base level of the steel bearing plate. The bearing plate should then be hammered into place to ensure that no void is left behind, and the nut tightened onto the soil nail reinforcement before application of the second stage shotcreting. Despite the promulgation of such good practice, honeycombing and/or voids underneath the bearing

Figure 7.19 Honeycombing and voids behind steel bearing plate of a soil nail head.

plate are observed during site audits from time to time (Figure 7.19). This emphasizes the importance of ensuring good construction practice on site and adequate site supervision by experience staff.

7.8.5 Lack of working space

In most cases, soil nail installation requires the erection of a working platform. If the working space is very limited, small and handy drilling rigs, such as those shown in Figures 7.3 and 7.4, could be used. Under special circumstances, the erection of a working platform for soil nailing works may not be feasible. For example, some busy roads could not be closed for more than a few hours during daytime whereby the associated roadside slopes may require upgrading to the required safety standards. In this case, erection of a working platform is not feasible. To overcome the site constraint, mobile platforms as shown in Figure 7.20 may be used, which can be mobilized quickly and removed at the end of work every day.

7.9 CONSTRUCTION SUPERVISION

7.9.1 General

It is important to supervise, inspect, and keep accurate and comprehensive records of all the construction activities in association with soil nailing

Figure 7.20 Mobile platform for soil nailing works on roadside slopes.

works because once the soil nails are installed, their quality is not readily visible. In general, full-time site supervision should be provided to all soil nailing works with certain critical activities, such as pullout tests, insertion of soil nail reinforcement, and grouting, to be individually inspected and checked by the site supervisory staff. The key aspects of supervision during soil nailing works are summarized in Table 7.2.

7.9.2 Supervision personnel

All soil nailing works should be supervised by personnel with suitable qualifications and experience. The site supervision team should be provided

Table 7.2 Key aspects of supervision during soil nailing works

Key Aspects	Supervision Items
Materials	• Undertaking inspections of soil nail reinforcement and its accessories to check on size, grade, length, corrosion protection measures, and integrity, and to ensure that all the soil nail components are assembled to the requirements of the specification.
Installation	• Checking to ensure that the soil nails, in particular their length, inclination, and spacing, are constructed in accordance with the design. • Monitoring and keeping records of the installation and testing operations of the soil nails, in particular the ground and groundwater conditions encountered during drilling, the volume of grout intake at a sustained low pressure head, and the process of pullout tests.
Buildability/ Safety	• Assessing the safety and adequacy of the methods used in constructing the soil nails, and the construction sequence, in particular for an excavation supported by soil nails. • Assessing the safety of temporary works and the effects of such works on the slope or the excavation, and the nearby ground, structures, facilities, and services. • Identifying any non-compliance with the specification or agreed method statements for temporary or permanent works and ensuring the situation is rectified promptly.

with sufficient information and briefing so that they can better appreciate the scope of works, key design assumptions, and the range of potential anomalies that can be encountered. The responsible supervision personnel should ensure that they are familiar with the relevant items including the site conditions and constraints, material requirements, construction drawings, specifications and testing requirements, and construction sequence. Suitable training and briefing should be given to the supervision personnel before the commencement of works.

7.9.3 Quality assurance

A quality assurance system should be established and implemented to ensure the quality of soil nailing works. As a minimum, the following key items should be included in the system:

- Appropriate construction materials are provided, and works are carried out in accordance with the construction drawings, method statement, and specification requirements.
- Allowable excavation heights are not exceeded.
- Setting out of soil nails is carried out to confirm every soil nail location on site.
- Drillholes are of correct size, length, inclination, orientation, and do not collapse during drilling and nail installation.
- Nail reinforcement is of correct size, length, and grade.

- Appropriate corrosion protection measures are specifically provided.
- Grouting and facing are in compliance with requirements.
- Appropriate surface, subsurface, and face drainage systems are provided.
- Appropriate testing, such as pullout test and materials compliance tests, is carried out as per specification requirements.

A sample checklist providing general questions that may need to be addressed when supervising the construction of soil nails is given in Table 7.3. The checklist should be modified to suit individual situations and contract requirements as appropriate.

7.10 QUALITY CONTROL TESTING

7.10.1 Materials compliance testing

All the materials used for the construction of soil nails should comply with the design and specification requirements. These include the soil nail reinforcement and the associated accessories, grout, heads and facing, and drainage elements. Materials compliance tests should be carried out on representative samples to verify the material quality. For cement grout, tests on cube crushing strength, bleeding, and flow cone efflux time are required. For soil nail reinforcement, tensile tests, bend tests, and re-bend tests should be carried out.

7.10.2 Pullout test

Field pullout tests carried out during the construction stage are classified as the verification/suitability test or acceptance/proof test (see also Section 6.4). The primary objectives of field pullout test are: (i) to verify the design assumptions on the bond strength at the interface between the ground, the reinforcement, and the cement grout annulus, (ii) to check the contractor's workmanship, (iii) to identify potential construction difficulties, and (iv) to ascertain the performance of soil nail without excessive deformation under working load condition.

In general, the verification/suitability pullout test is to verify the compliance of the design pullout capacity based on the contractor's proposed installation method. This type of pullout test should, as far as practicable, be carried out at locations where the pullout resistance may be low or the buildability of the soil nails is fraught with difficulty, e.g. at locations of relatively weaker material or high groundwater level. Moreover, the test locations should be determined in such a manner that at least two tests are conducted in the anticipated major load-bearing soil stratum. Verification pullout tests should be carried out prior to the construction of working

Table 7.3 Sample checklist for soil nail construction control

	Activities	Findings			
No.	Description	Yes	No	N/A	Remarks
1	**Pre-Construction Review**				
1.1	Any approved drawings, geotechnical design reports, and specification?				
1.2	Any approved method statements detailing construction procedures and sequences of works?				
1.3	Any material requirements, construction tolerances, and acceptance/rejection criteria?				
1.4	Any compliance testing requirements to ensure the quality of the works?				
1.5	Any monitoring requirements to check the performance of the works?				
1.6	Any temporary works required to facilitate the construction of the permanent works?				
1.7	Any pre-construction site trial to assess the buildability of the works?				
2	**Setting Out**				
2.1	Are the positions of the soil nails in agreement with the contract requirements?				
2.2	Have the positions of the soil nails been checked to see whether any existing utilities, channels, surface boulders, trees, foundations, and other structures or any proposed works such as surface channel and subsurface drains would be affected?				
3	**Drilling**				
3.1	Has the drilling equipment (type, diameter of drill bit, total length of drill rods, flushing medium, etc.) been checked?				
3.2	Has water, dust, fumes, and noise generated during the drilling operation been sufficiently diverted, controlled, suppressed, or muffled?				
3.3	Have the drilling works on working platforms, which are visible to nearby residents, been shielded from view by tarpaulin sheets?				
3.4	Any requirement on use of permanent or temporary casing?				
3.5	Are there any freshly grouted soil nails near the drillhole to be drilled?				
3.6	Are the drillhole diameter, length, inclination, and bearing in accordance with the contract requirements?				

(Continued)

Table 7.3 (Continued) Sample checklist for soil nail construction control

	Activities	Findings			
No.	Description	Yes	No	N/A	Remarks
3.7	Has the Contractor suitably controlled the drill rate to minimize the eccentricity produced by the dip of the drill rods when drilling long soil nails?				
3.8	Have random checks been carried out on inclination and bearing of the drillhole during drilling?				
3.9	Are there any anomalies in respect of the ground and groundwater conditions?				
4	**Assembly of Soil Nail Reinforcement**				
4.1	Are the soil nail components, including reinforcement, grout pipes, centralizers, couplers, corrugated plastic sheathing, heat-shrinkable sleeve, washers, nuts, bearing plates, and conducting wires (for NDT, see Section 7.10.4) of the correct type, grade, length, and size?				
4.2	Are the centralizers adequate to support the reinforcement and ensure minimum grout cover?				
4.3	Have the couplers been inspected for tightness after assembly?				
4.4	Are the grout pipes straight, free from blockage, without side holes (except near the end of the pipe as specified in the contract) and extended to the end of the soil nails?				
4.5	Are the corrosion protection measures to reinforcement and reinforcement connectors in accordance with the contract requirements and have they been inspected for integrity?				
4.6	Has the assembling method been verified by site trials for not causing damage, deformation, and displacement to the soil nail components on completion of assembly, during insertion, and withdrawal of the soil nails?				
5	**Installation**				
5.1	Have the drillholes been left open for a time longer than that permitted in the contract?				
5.2	Is there any constant flow of water coming out from the drillhole?				

(Continued)

Table 7.3 (Continued) Sample checklist for soil nail construction control

	Activities	Findings			
No.	Description	Yes	No	N/A	Remarks

5.3	Have the drillholes been cleared of debris and standing water immediately before installation of reinforcement?				
5.4	Has the correct assembly of the soil nail reinforcement been inserted?				
5.5	Have obstructions been encountered during insertion of reinforcement into the drillhole?				
5.6	Has a minimum clearance, in accordance with the contract requirements, been maintained between the distal end of the reinforcement and the bottom of the drillholes after installation?				
6	**Grouting**				
6.1	Is the grout mix in accordance with the contract requirements?				
6.2	Has the grouting operation been carried out in accordance with the method statement?				
6.3	Has the grouting of soil nails been carried out on the same day as the soil nail installation?				
6.4	Have water, dust, fumes, and noise generated during the grouting operation been sufficiently diverted, controlled, suppressed, or muffled?				
6.5	Is there any excessive grout take?				
6.6	Is the grout that has returned from the top of the drillhole of satisfactory cleanliness and viscosity?				
6.7	Has a minimum pressure head in accordance with the contract requirements been maintained in the outlet pipe after completion of grouting until the cement grout has reached the initial set?				
7	**Construction of Soil Nail Heads**				
7.1	Are the soil nail heads of correct size and the materials in accordance with the contract requirements?				
7.2	Have the threads at the proximal end of reinforcement been thoroughly cleaned, properly treated with hot-dip galvanized coating, or protected with approved zinc-rich paint prior to construction of soil nail heads?				

(Continued)

Table 7.3 (Continued) Sample checklist for soil nail construction control

No.	Activities	Findings			Remarks
	Description	Yes	No	N/A	
7.3	Has the placed concrete been adequately compacted to avoid honeycombing?				
	Has the concreting of the soil nail heads been divided into two stages when using sprayed concrete?				
	Have water, dust, fumes, and noise generated during the concreting operation been sufficiently diverted, controlled, suppressed, or muffled?				
	Are there any anomalies in the workmanship of the soil nail heads which have been uncovered?				
	Has the corrugated plastic sheathing been embedded into the soil nail heads in accordance with the contract requirements?				
	Pullout Test				
	Have adequate test soil nails been installed for the pullout test?				
	Has the pullout test equipment been set up in accordance with the contract requirements?				
	Any need to carry out a creep test?				
	Have the test soil nail drillholes been fully grouted after completion of the tests?				
	Excavation				
	Are the excavation and the soil nail construction sequence in accordance with the method statement?				
	Have the soil nails and soil nail heads been constructed in time?				
	Has the excavation surface been protected from water ingress and surface erosion?				
	Are the temporary drainage provisions adequate?				
	Any excessive movement affecting the stability of the excavation or nearby facilities?				
	Site Supervision				
	Has the required qualified supervision been provided, in particular at the critical stages of the soil nailing works?				

nails so that the information gathered from the tests can be reviewed for making design changes as necessary. If the contractor makes major changes to the installation method, or if significant variations in the ground conditions are observed, additional verification/suitability pullout tests may be required.

The soil nails to be tested should be sacrificial and installed using the same construction procedures as the working nails, except that only the bottom part of the soil nail needs to be grouted. In Hong Kong, the length of the cement grout annulus of test nail is typically 2 m. Too small a bond length may not be representative, whereas a large bond length requires a large pullout load and hence heavy equipment and set-up. Nevertheless, designers may choose to specify a grouted length other than 2 m to suit particular test objectives. Verification pullout test should be conducted to failure or, as a minimum, to a test load that corresponds to the design bond strength times the required factor of safety against pullout failure. Typical procedures and acceptance criteria of the test as adopted in Hong Kong are given in Figure 6.5 of Chapter 6.

Soil nails subject to verification pullout tests require partial grouting of the drillholes to form the specified bond length for testing. Grouting should be carried out slowly and carefully to prevent over-grouting. Packers are usually used to seal off the grouted section. Many types of packers, such as inflatable packers, are available. Only those packers that can effectively seal off the grouted section should be used (e.g. Figure 7.21 shows the details of packers commonly used in Hong Kong). The packers should, as far as practicable, not contribute to the bond strength of the grouted section, or else any such contribution should be taken into account in the estimation

Figure 7.21 Fixing details of a packer in Hong Kong.

of bond strength of the soil nail. Recently, the time domain reflectometry (TDR) technique has been used as an alternative to the use of packer to determine the length of the grouted section with reasonable accuracy (see Section 7.10.4 for more details).

Acceptance/proof pullout tests are intended to verify the consistency of soil nail performance. These pullout tests should be carried out on sacrificial or production nails during or on completion of the construction works. The test should be conducted where issues were observed during installation of selected production nails, such as irregular drilling progress, lower than typical grout take, or those nails that had to be re-drilled. If production nails are load tested, caution should be exercised to avoid overstressing the reinforcement-grout bond and damaging the integrity of the grout and the corrosion protection measures. In the United States, about 5% of the total production nails are load tested to 75% of the design load, whereas there is no such requirement in prevailing practice in Hong Kong because of the above concerns about possible adverse effect on the production nails. It is good practice to collect information on the type of material encountered and the presence of groundwater during hole drilling in order to learn more about the site-specific ground conditions.

7.10.3 Creep test

For soil nails designed to carry sustained loads and are bonded in soil, a creep test should be carried out to determine the susceptibility of the soil nail to long-term creep. The test can be conducted as part of the verification and acceptance pullout test. The number of creep tests may be the same as that for pullout tests. Nevertheless, one should exercise judgment regarding whether the creep tests are sufficient and representative in meeting the test objective. A creep test consists of measuring the movement of the test nail at a constant load over a specified period of time. Typical procedures and acceptance criteria of the test as adopted in Hong Kong are given in Figure 6.6 of Chapter 6.

In the event that the acceptance criteria cannot be met by the creep test, the design bond strength of the soil nail should be reviewed and revised as appropriate. In some situations, the bond zone may have to be relocated to a different geological material in order to achieve the required bond strength. Further creep tests should be carried out if the assumption on the design bond strength has changed.

7.10.4 Non-destructive testing

7.10.4.1 General

Soil nails have been used extensively in earth structures since the 1980s. However, once a soil nail has been installed, it is difficult to check the quality of the burial works, such as the length of the steel bar and the integrity of cement grout annulus. Non-destructive testing (NDT) is an attractive means which encourages higher construction standards and promotes improvements in installation techniques and better quality control and workmanship. The test results can also be used to build up an overall picture of the integrity of the soil nailed structure.

The Geotechnical Engineering Office in Hong Kong began in 2001 to identify and study various NDT methods that could help to assess the quality of the installed soil nails in respect of length of the reinforcement and integrity of the cement grout annulus. The NDT methods are to provide additional quality assurance and serve as a deterrent against malpractice. Among the potential NDT methods examined, which included the sonic echo method, the *mise-a-la-masse* method, the electro-magnetic induction method, the electrical resistance method, magnetometry, and time domain reflectometry (TDR), TDR was found to be a simple, reliable, quick, and inexpensive method for the above objectives (Cheung, 2003; Cheung & Lo, 2011). The principle and application of the TDR technique to soil nailing are given in the following.

7.10.4.2 Principle of TDR

TDR was first developed in the 1950s based on the principle of radar. Instead of transmitting a three-dimensional wave front in radar, the propagation of the electromagnetic wave in TDR is confined in a waveguide (O'Connor & Dowding, 1999). TDR is commonly used in the telecommunication industry for the identification of discontinuities in transmission lines. In the 1980s, the application of the technique was extended to many other areas, such as geotechnology, hydrology, and material testing (e.g. Dowding & Huang, 1994; O'Connor & Dowding, 1999; Siddiqui et al., 2000; Liu et al., 2002; Lin & Tang, 2005). In theory, TDR involves sending electrical pulses along a transmission line, which is in the form of coaxial or twin-conductor configuration, and receiving reflections or echoes induced by any discontinuities or mismatches in electrical properties in the transmission line. By measuring the time for the pulses to travel from the pulse generator to the point of discontinuity or mismatch, the corresponding distance can be estimated using Equation (7.1) if the pulse propagation velocity, v_p, is known.

$$L = v_p t \tag{7.1}$$

where L is the distance between the pulse generator and the point of discontinuity or mismatch, and t is the respective pulse travel time. The pulse propagation velocity, v_p, is related to the electrical properties of the material in close proximity to the pair of conductors by the following expression (Topp et al., 1980):

$$v_p = \frac{v_c}{\sqrt{\epsilon}} \tag{7.2}$$

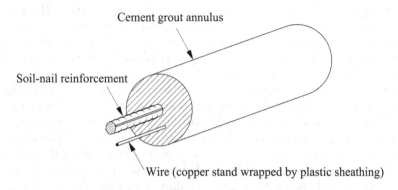

Figure 7.22 Analogy of a soil nail with pre-installed wire as a twin-conductor transmission line.

where v_c is the speed of light in vacuum (3×10^8 m/s) and ε is the dielectric constant which measures how a material reacts under a steady-state electric field (for air, $\varepsilon \approx 1$, and for cement grout, $\varepsilon \approx 10$).

If a wire is pre-installed alongside a soil nail reinforcement, which is generally a steel bar, as shown in Figure 7.22, the configuration becomes analogous to a twin-conductor transmission line and the end of the grouted reinforcement-wire pair becomes a discontinuity. This suggests that TDR can be used to determine the length of the installed soil nails (Cheung, 2003).

As indicated in Equation (7.1), one can estimate the length of an installed soil nail pre-installed with a wire if the two key parameters, namely the time for a pulse to travel from the soil nail head to its end, t, and the pulse propagation velocity, v_p, are known. Equation (7.2) further suggests that the pulse propagation velocity, v_p, along a reinforcement-wire pair in air will be much higher (about two to three times) than that in cement grout. Hence, the pulse travel time along a soil nail with voids in the grout annulus will be less than that in a fully grouted soil nail of the same length. In addition, a reflection will be induced at the grout-void interface.

The magnitude and polarity of the reflection when an electrical pulse reaches the location of discontinuity or mismatch will depend on the amount of changes in electrical impedance at this location, which can be expressed in terms of the reflection coefficient, Γ (Hewlett Packard, 1998):

$$\Gamma = \frac{V_r}{V_i} = \frac{Z - Z_o}{Z + Z_o} \tag{7.3}$$

where V_r is the peak voltage of the reflected pulse, V_i is the peak voltage of the incident pulse, Z_o is the characteristic electrical impedance, and Z is the electrical impedance at the point of reflection.

Based on Equation (7.3), one may infer the possible waveforms for different soil nail configurations. For example, when a pulse reaches the end of a soil nail, a positive Γ (i.e. positive pulse reflection) will occur due to an increase in electrical impedance (i.e. Z_o (grouted reinforcement-wire pair) $< Z$ (pair end)). However, if the wire is in electrical contact with the reinforcement at the end of a soil nail, Z (pair end) will tend to 0 and Γ becomes negative (i.e. negative pulse reflection). Similarly, a positive reflection will be obtained when a pulse passes from cement grout to void due to an increase in electrical impedance, whereas a negative reflection will be returned when the pulse passes from void to cement grout. Thus, the waveform manifests as an S-shaped reflection when the pulse passes through a voided section in the grout annulus.

Figure 7.23 shows schematically some basic TDR waveforms for a standard transmission line and the corresponding soil nail with different configurations.

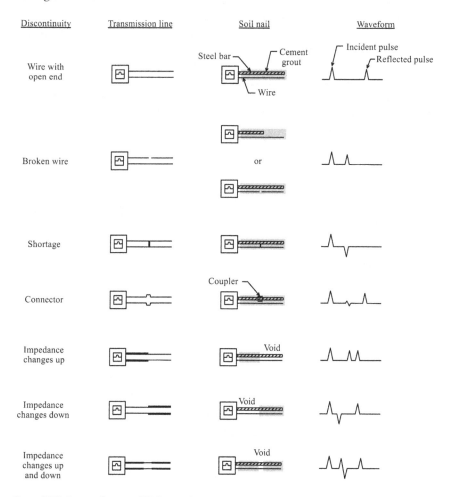

Figure 7.23 Basic schematic TDR waveform of a transmission line and an analogous soil nail.

7.10.4.3 Control test

In order to evaluate the feasibility of using TDR to check the quality of installed soil nails in respect of length and integrity of cement grout annulus, control tests were carried out on some soil nails of various known lengths (from 4 m to 15 m) and of various known grout annulus configurations (void size varies from 0.2 m to 3 m in length). These soil nails were prefabricated under factory conditions so that the dimensions of the void, grout annulus, length, etc., can be controlled at a reasonably accurate level. The diameters of the reinforcement, wire and cement grout annulus are 25 mm, 2 mm, and 100 mm, respectively. The wire was fixed alongside a reinforcement by plastic tape at 2 m intervals. The testing device is a commercial cable fault detector available in the market with an accuracy of more than 99.5%. Figure 7.24 shows the TDR waveforms obtained from fully grouted soil nails of different lengths. Positive reflections are returned from the respective ends and the time of pulse propagation is proportional to the length of the soil nails.

Figure 7.25 shows a comparison between the waveform obtained from a fully grouted soil nail and that from a soil nail of the same length but with intermittent void sections. It is evident that when the pulses pass the locations of the void section, as indicated in Figure 7.25b, there are S-shaped

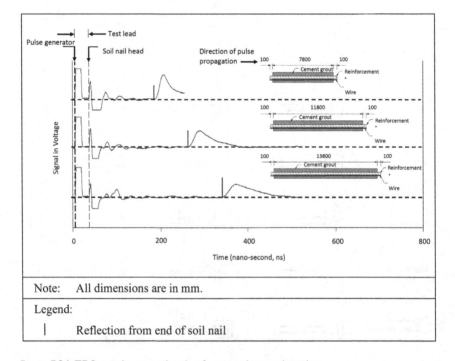

Figure 7.24 TDR results on soil nails of various known lengths.

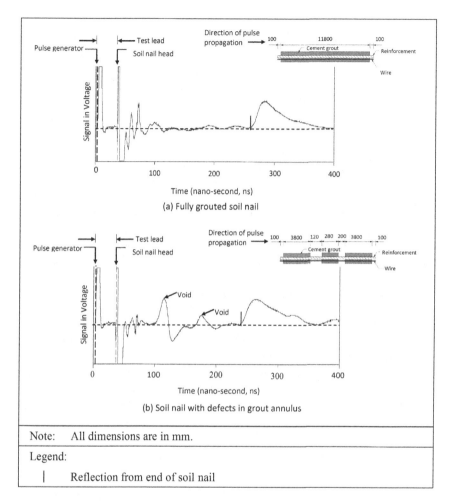

Figure 7.25 TDR results on a fully grouted soil nail and a soil nail with defects in grout annulus.

reflections. One may notice that if the two prefabricated void sections are placed close to each other, the two S-shaped reflections will be overlapped, and it may become difficult to identify their relative locations. Both the fully grouted soil nail and the soil nail with void sections returned a positive reflection at their ends. However, the pulse travel time in the soil nail with void sections is slightly less than that in the fully grouted nail. The control tests suggest that it is feasible to use TDR to infer the length of a soil nail as well as the integrity of the cement grout annulus. The possible sources and degree of uncertainty associated with the technique, however, need to be examined prior to field application.

7.10.4.4 Accuracy of TDR test

By nature, NDT is an indirect way of measuring physical objects where it is inevitable that uncertainty will be introduced during the course of measurement and interpretation of test results, and thus it will not give definitive answers. A TDR test basically comprises two parts, viz. determination of pulse propagation velocity from soil nails of known lengths and grout integrity (i.e. calibration), and measurement of the time for a pulse to propagate from the head of a test soil nail to its end and recognition of any anomaly in the waveform (i.e. testing). Any factors that affect the two key parameters in Equation (7.1) (i.e. the pulse propagation velocity, v_p, and the pulse propagation time, t) will introduce uncertainty to the TDR-deduced length. The uncertainty of a TDR test can be broadly divided into two categories, namely nail-independent and nail-dependent uncertainties. As shown in Figure 7.26, amongst the sources of uncertainty, those involving built-in error of testing instrument and human judgment are classified as nail-independent uncertainty, while the remaining sources are classified as nail-dependent uncertainty. The former uncertainty is not affected by the variability in soil nail configuration and properties (e.g. reinforcement size, wire type, grout integrity, etc.), while the latter includes the variability in the soil nail configuration and properties, and other unaccounted factors.

The uncertainties associated with a TDR test and the inherent natural variation of quality of soil nails have been identified and assessed (Cheung, 2006) in order to facilitate the determination of the precision limit of the test method (i.e. test-related uncertainty) and the effect of inherent normal variation of soil nails on the test results (i.e. test-unrelated uncertainty). Amongst the various sources of uncertainty in soil nail length estimation,

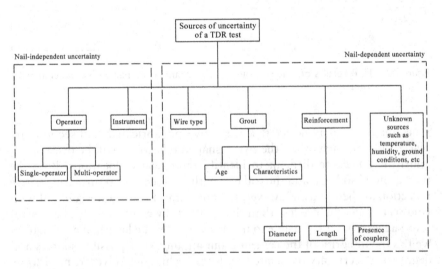

Figure 7.26 Sources of uncertainty of a TDR test.

some of them can be minimized. For example, the variability associated with the difference in wire type can be minimized by using the same type of wire in the calibration and test soil nails. The uncertainty due to human judgment can be reduced by standardizing the testing procedure and provision of guidelines on result interpretation. Similarly, the effect of variability of grout characteristics amongst soil nails can be minimized by determining the pulse propagation velocity from soil nails installed on the same construction site as the test soil nails. Nevertheless, irrespective of the sources of uncertainty, the maximum error bound can be estimated when the estimated lengths are compared with the as-built lengths of the test soil nails without making any distinction with respect to operators, reinforcement diameter and length, wire type, characteristics of grout, etc. In this regard, a sufficiently large pool of 700 test soil nails spreading over nine different construction sites was used for the exercise. The frequency distribution of the length difference is presented in Figure 7.27a. It is noteworthy that the number of measurements is greater than 700 because some of the soil nails were tested by four operators for the estimation of multi-operator uncertainty. The data set is observed to fit a normal distribution. The Kolmogorov-Smirnov goodness-of-fit test for the assumed normal distribution is not rejected at the 5% significance level. By fitting the probability distribution of the length difference with a normal model, as in Figure 7.27b, the mean and standard deviation of the model are found to be 0.82% and 4.16%, respectively. This suggests that there is a 95% confidence level that the difference in length between the estimated value and as-built value due to overall uncertainty falls within −7.3% and +8.9% (i.e. 0.82% ± 1.96 × 4.16%) of the as-built length. Amongst the sources of uncertainty, some 6.6% is due to built-in error of instrument and human judgment (i.e. nail-independent uncertainty). This suggests that the variability of the configuration and properties of these 700 soil nails is not substantial.

Based on the pool of test results, the 95% confidence level of the overall error in length estimation using TDR is estimated to be about ±9%. This includes both the nail-dependent and nail-independent uncertainties. Apart from the uncertainty of the TDR test method, consideration was given to the number of defective nails to be detected against the number of false alarms. As a result, an "alert limit" was devised at ±15% of the design length such that if the difference between the TDR-deduced length of a soil nail and its design length exceeds the alert limit, that soil nail is considered as anomalous and follow-up action will be initiated. A short TDR-deduced length could be due to either the as-built length of the reinforcement, or the pre-installed wire, or both, being shorter than their corresponding design length and/or there are substantial defects in the grout sleeve. In addition to the deduced length, the TDR waveforms could also provide some telltale signs of certain anomalies. Thus, an anomalous test result can be a short TDR-deduced length, or a short TDR-deduced length coupled with an anomalous TDR waveform.

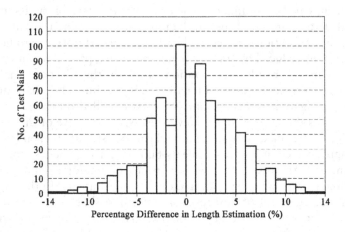

(a) Frequency Distribution of Estimation Error

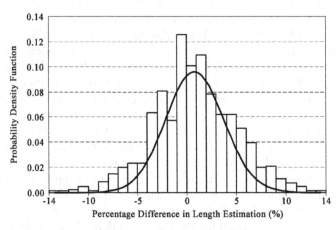

(b) Probability Density Function of Estimation Error

Note:	% Difference in length estimation	=	Estimated nail length - As-built nail length
			As-built nail length

Figure 7.27 Overall uncertainty in soil nail length estimation using TDR test.

7.10.4.5 Case history

Since 2004, TDR has been widely used in Hong Kong as part of the quality check during independent site audits of soil nailing works. Up to 2018, more than 10,000 soil nails have been tested using TDR. In general, the percentage difference between the TDR-deduced length and design length of most of the tested soil nails, as shown in Figure 7.28, does not exceed the alert limit. There are, however, a small number of soil nails (less than

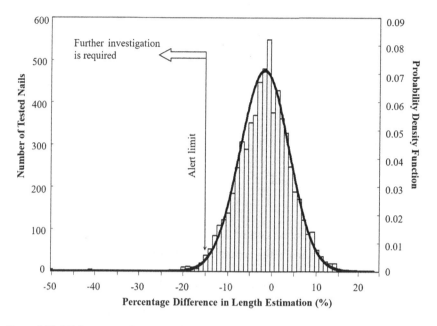

Figure 7.28 TDR test results in the quality assurance program.

1%) with such difference exceeding the alert limit and further investigation was conducted. Most of the soil nails with short TDR-deduced length are isolated cases, i.e. an anomaly found in only one or two soil nails at a site and additional TDR tests on adjacent soil nails do not show any anomaly. Generally, a design review would be carried out assuming the anomalous nail to be not fully functional.

TDR is found to be an effective tool to supplement site supervision of soil nailing works, whose quality cannot be checked easily after construction. With the help of TDR, the anomalies in the soil nailing works could be identified and follow-up actions taken promptly. More details about the engagement of TDR in the quality control of soil nailing works can be found in Cheung (2006) and Cheung & Lo (2011). In the following, a case is presented in which TDR has identified successfully the anomalous soil nails such that appropriate remedial actions were taken swiftly.

An existing cut slope of about 10 m high and 115 m long with an average slope angle of 50° was reinforced by soil nails. About 100 soil nails of 7 m long were installed into the slope by the "drill-and-grout" method. During an independent site audit, ten soil nails were selected randomly for the TDR test. Amongst the ten tested nails, five were found to have TDR-deduced length significantly shorter than their design length and the corresponding TDR waveforms were anomalous. Subsequently, additional TDR tests were carried out on all the remaining soil nails at the slope and two more soil nails were revealed to have short TDR-deduced length and anomalous TDR

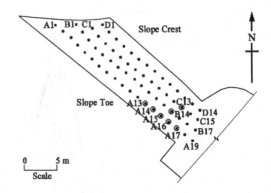

Legend:

    •   Soil nails with normal TDR test results

    ◉   Soil nails with anomalous TDR test results

Figure 7.29 Layout of soil nails at northern portion of the slope.

waveform. The seven soil nails with anomalous TDR test results, namely A13 to A17, B14, and C13, clustered at the northern portion of the slope (see Figure 7.29). The TDR waveforms of these seven soil nails differ from those of the remaining soil nails in two aspects as illustrated in Figure 7.30, namely (i) the presence of significant local reflections preceding the major reflection from the nail end, and (ii) shorter pulse travel time to the end of these soil nails.

The waveforms of the seven soil nails (e.g. Figure 7.30b) bear some resemblance to the characteristics of that with grout defects in Figure 7.25b, suggesting the anomalies in these seven soil nails could be related to the presence of voids in the cement grout annulus. A review of the site records also indicated that significant grout take was encountered at these locations during the construction of the soil nails.

Given that the sizes of the void inferred from TDR results could be substantial, it was decided to remove the concrete nail heads for inspection. Upon removal of the concrete nail heads of the seven nails, void sections were noted. Subsequently, seven 150 mm diameter holes were sunk vertically in the vicinity of the concerned area and a digital camcorder was lowered through these holes for inspection. A cavity with a size of about 15 m^3 situated at about 6 m to 8 m below the ground level was uncovered. The cavity was subsequently filled with concrete and the seven soil nails were replaced. TDR was used to check the grout integrity of the replacement soil nails both during grouting operation and after installation. The case history presented indicates that TDR can play a useful role in assuring construction quality and promoting improvements in installation techniques and quality control of soil nailing works.

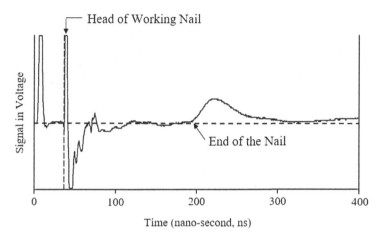

(a) Normal TDR waveform

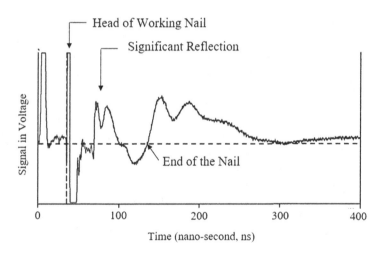

(b) Anomalous TDR waveform

Figure 7.30 Normal and anomalous TDR waveforms.

7.10.4.6 *Application of TDR in other areas related to soil nailing works*

Apart from quality control of soil nailing works, TDR can also be used as an alternative to the use of packers to determine the grouted length of soil nails for pullout test. This involves installation of a wire with a length that equals the free length of the soil nail, L_U (i.e. $L_U = L - L_G$) (see Figure 7.31). Following the principle of TDR described above and the basic TDR waveforms in Figure 7.23, it is anticipated that a positive

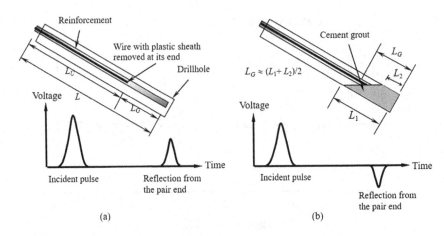

Figure 7.31 Schematic TDR waveforms of soil nail before and after grouting.

reflection will be returned at the end of the reinforcement-wire pair (i.e. end of the wire) prior to grouting works (Figure 7.31a). When the level of grout touches the end of the wire, the wire and the reinforcement will be electrically in contact via the fluid grout. Consequently, the polarity of the reflected pulse will switch from positive to negative (Figure 7.31b). At this incidence, the grouting operation can be stopped. The as-built grouted section will have a shape and length similar to that shown in Figure 7.31b. Field trials were carried out in Hong Kong and the results indicated that the TDR technique could be used to determine the as-built average grouted length, L_G, of soil nails in pullout test with a fairly high accuracy. A limitation of the technique is that it may not be feasible if the grouted zone is submerged in groundwater, in particular when the groundwater contains high mineral content, where a close circuit will be formed regardless of the presence of cement grout. For more details about the use of TDR in determination of grouted length, please refer to Cheung et al. (2008).

7.10.5 Destructive testing

Destructive testing of soil nails is seldom feasible. Where needed, destructive testing techniques, such as stitch drilling, over-coring, and excavation, are occasionally adopted to exhume the installed soil nails to check their lengths and as-built condition (Figure 7.32). However, it can be difficult to control the alignment of drilling and coring with respect to that of the installed soil nail. Excavation to expose long soil nails may even be impractical.

Figure 7.32 Exhumation of soil nail using stitch drilling.

7.11 CONCLUDING REMARKS

Proper construction is vital in ensuring satisfactory performance of a soil nailed structure. In this chapter, the key aspects during the construction stage concerning the safety and quality of soil nailed structures are identified and discussed. These include various soil nail installation methods, construction sequence, and the key elements of the quality assurance system. Difficulties that may be encountered during the construction of soil nails, together with the corresponding preventive and mitigation measures, are also highlighted. It is of paramount importance that designers, contractors, and site supervision teams fully understand the merits and limitations of the soil nailing technique, the key design assumptions, site settings, scope of works, and the possible challenges that may be encountered during construction.

REFERENCES

Cheung, W.M. (2003). Non-Destructive Tests for Determining the Lengths of Installed Steel Soil Nail. Geotechnical Engineering Office, Civil Engineering and Development Department, HKSAR Government, Hong Kong, GEO Report No. 133, 54 p.

Cheung, W.M. (2006). Use of Time Domain Reflectometry to Determine the Length of Steel Soil Nails with Pre-installed Wires. Geotechnical Engineering Office, Civil Engineering and Development Department, HKSAR Government, Hong Kong, GEO Report No. 198, 35 p.

Cheung, W.M., Lo, D.O.K. & Pun, W.K. (2008). Use of time domain reflectometry in soil nailing works. In Proceedings of the HKIE Geotechnical Division Annual Seminar. The Hong Kong Institution of Engineers, Hong Kong, pp. 89–94.

Cheung, R.W.M. & Lo, D.O.K. (2011). Use of time-domain reflectometry for quality control of soil nailing works. *Journal of Geotechnical and Geoenvironmental Engineering, ASCE*, vol. 137, pp. 1222–1235.

Dowding, C.H. & Huang, F.C. (1994). Early detection of rock movement with time domain reflectometry. *Journal of Geotechnical Engineering, ASCE*, vol. 120, pp. 1413–1427.

Hewlett Packard (1998). *Time Domain Reflectometry Theory (Application Note 1304-2)*. Hewlett Packard Company, USA, 16 p.

HKSAR (Hong Kong Special Administrative Region) (2013a). *Guidance Notes of Design for Safety*. Development Bureau, HKSAR Government, Hong Kong, 68 p.

HKSAR (2013b). *Worked Examples of Design for Safety*. Development Bureau, HKSAR Government, Hong Kong, 171 p.

O'Connor, K.M. & Dowding, C.H. (1999). *GeoMeasurements by Pulsing TDR Cables and Probes*. CRC Press, 394 p.

Lin, C.P. & Tang, S.H. (2005). Development and calibration of a TDR extensometer for geotechnical monitoring. *Geotechnical Testing Journal, ASTM*, vol. 28, no. 5, pp. 464–471.

Liu, W., Hunsperger, R., Chajes, M., Folliard, K. & Kunz, E. (2002). Corrosion detection of steel cables using time domain reflectometry. *Journal of Materials in Civil Engineering, ASCE*, vol. 14, no. 3, pp. 217–112.

Lui, B.L.S. (2009). Soil Nail Hole Measurements (GEO Report No. 245). Geotechnical Engineering Office, Civil Engineering and Development Department, HKSAR Government, 98 p.

Siddiqui, S.I., Drnevich, V.P. & Deschamps, R.J. (2000). Time domain reflectometry for use in geotechnical engineering. *Geotechnical Testing Journal, ASTM*, vol. 23, no. 1, pp. 9–20.

Topp, G.C., Davis, J.L. & Annan, A.P. (1980). Electromagnetic determination of soil water content: Measurement in coaxial transmission lines. *Water Resources Research*, vol. 16, no. 3, pp. 574–582.

Yue, Z.Q., Lee, C.F., Law, K.T. & Tham, L.G. (2004). Automatic monitoring of rotary-percussive drilling for ground characterization - illustrated by a case example in Hong Kong. *International Journal of Rock Mechanics and Mining Science*, vol. 41, pp. 573–612.

Chapter 8

Monitoring and maintenance

8.1 INTRODUCTION

Similar to other civil engineering and building works, monitoring and maintenance are generally essential to assess and upkeep the short-term or long-term performance of a soil nailed structure. Monitoring activities are usually carried out to assess the short-term performance, mostly during and shortly after construction, of a soil nailed structure. Long-term monitoring is, however, not required in general unless there is a specific purpose for such requirement, such as for research and technical development. Maintenance is generally required to upkeep both the short-term and long-term performance of soil nailed structures. The information collected and experience gained through monitoring and maintenance also helps to enhance the design process. This chapter provides an overall view of the issues in relation to the monitoring and maintenance of soil nailed structures. Two cases are presented to illustrate the implementation of monitoring schemes with deployment of various types of instruments at two construction sites of different characteristics. However, it is out of scope of this chapter to provide a full coverage of all the instrumentation aspects and readers are recommended to refer to other publications such as Dunnicliff (1993), Slope Indicator (2004), and Transport for NSW (2016) for more details in this area.

8.2 MONITORING

8.2.1 Monitoring plan

The most representative parameter in reflecting the performance of a soil nailed structure is the amount of deformation experienced by the structure during and after construction. Although many thousands of soil nailed structures have been constructed in different parts of the world over the past decades, only a limited number have been instrumented with monitoring devices to provide long-term behavior and performance information of these structures. Instead, it is more common that monitoring of a nailed structure is carried out during or shortly after its construction, in particular when there are existing facilities that may affect or be affected by the construction activities. Generally speaking, monitoring is undertaken for the following reasons:

- Verification of design parameters—Information collected is used to verify the predicted versus actual behavior of the nailed structure. It serves as a part of the observational method during the construction stage whereby design modification using revised parameters can be made if necessary.
- Confirmation of the safety level of the nailed structure and the facilities being potentially affected by the construction—Information collected can be used to assess the condition of the nailed structure and the facilities in the vicinity. If the parameter being monitored, such as ground deformations and loads in soil nails, exceeds the pre-set threshold value, the necessary mitigation or remedial actions may be taken.
- Collection of data for research purposes—Information collected can enhance our understanding of the fundamental principle governing the behavior of nailed structures. For example, long-term performance monitoring of nailed structures in earthquake-prone regions can provide insights into future seismic design of soil nails.

Readers should note that the necessity of a monitoring plan during or after the construction of a nailed structure should be determined at the early stage of the design. If affirmative, the monitoring plan should be developed as part of the detailed design. As a minimum, a monitoring plan should include the following elements:

- The purposes of monitoring.
- A geotechnical baseline in conjunction with a summary of the existing conditions of the site and the facilities to be monitored. Other relevant information, such as condition survey of existing structures, should also be included.
- A plan containing general information of the proposed soil nailed structure and the existing facilities in the vicinity of the site. The plan should also indicate the location, number, type, and details of the instrumentation. The reasons for selecting particular type of instrument should also be clearly spelled out.
- A list of parameters to be monitored and the corresponding frequency.
- A list of testing, site, or laboratory test, and the corresponding frequency.
- Predicted magnitudes of the parameters being monitored.
- Assessment on the performance of the soil nailed structure and the concerned facilities that may be affected by the construction. The

proposed mitigation or remedial actions to be taken should also be included if the parameters being monitored exceed the pre-set threshold values.

8.2.2 Instrumentation and parameters to be monitored

The key parameter representing the overall performance of a soil nailed structure is the deformation of the structure during and after construction. Other pertinent parameters include the magnitude and distribution of loads developed in the soil nails, nail heads, and facings, condition of the ground (e.g. soil corrosivity) and groundwater, condition of surface and subsurface drainage systems, temperature, rainfall, etc.

The selection of monitoring instruments should be based on the parameters to be measured, the reliability, precision, simplicity, and durability of the instruments, and the availability of the instruments in the market. A brief summary of the various types of common monitoring instruments for assessing performance of soil nailed structures is given in Table 8.1.

8.2.3 International practice

The practice of monitoring of soil nailed structures currently adopted by various places including France, Hong Kong, and the United States has been reviewed (Clouterre, 1991; Byrne et al., 1998; Lazarte et al., 2003, 2015; GEO, 2008; O'Donovan et al., 2020). It is a common practice that monitoring is carried out during the construction of a nailed structure. The parameters to be monitored are mostly related to ground deformation, deformation and signs of anomaly of the nailed structure, and the facilities being potentially affected by the construction activities, such as cracks and groundwater levels. Monitoring of the horizontal and vertical movement at the crest and facing of the nailed structure being constructed as well as the groundwater level is commonly carried out at all stages of construction to confirm its temporary stability. For example, as a minimum, monitoring during nailed wall construction in the United States will normally include the following:

- Vertical and horizontal movements of the nailed wall face using survey points and optical surveying methods.
- Ground deformation profile using inclinometers installed at a short distance, e.g. 1 m, behind the wall facing.
- Signs of disturbance in the ground surface and the wall face using crack gauges.
- Groundwater level fluctuation using piezometers installed behind the wall facing.

Table 8.1 Common types of instruments for assessing performance of soil nailed structures

Parameters to be Monitored	Instruments	Details
Subsurface Lateral Deformation	Inclinometers	• Inclinometers monitor the magnitude, direction, and rate of subsurface lateral deformations. They provide the most comprehensive information on the deformation of a nailed structure and the ground in the vicinity.
		• Inclinometers can be installed at different elevations of a soil nailed structure, so as to provide comprehensive deformation data. An inclinometer consists of a grooved casing grouted vertically in a borehole. The inclinometer casing must be installed into rock or dense material that is not expected to deform, which provides a point of fixity where other measurements through the casing can be reliably correlated to. The technology is well-established, and the instrument is readily available in the market. In general, inclinometers should be installed preferably about 1 m behind the crest of a soil nailed structure.
		• There are two types of inclinometers, namely portable traversing probe inclinometers and in-place inclinometers. Portable traversing probe inclinometers are not suitable for sites where continuous monitoring is required. Recently, there is an increasing trend of using in-place inclinometers, which enable rapid measurements at points throughout the full depth of the inclinometer without the need to physically introduce and remove a probe into the inclinometer casing to take readings at every depth. Nonetheless, in-place inclinometer readings should be calibrated with manual readings using a standard probe before application.
Settlement	Settlement Check Points	• Settlement check points can be installed on nailed structures and other concerned facilities. They can be in the form of pins driven into the ground or mounted on a structure, or simply be a painted reference point on a structure. These check points are monitored using conventional surveying methods.
	Horizontal Inclinometers (Horizontal Profile Gauge)	• The horizontal inclinometers can be installed at different elevations of a soil nailed structure. The inclinometer casing is installed horizontally and attached to a remote pulley unit. The device consists of a traversing probe, a control cable, pull cable, inclinometer casing, and read

(Continued)

Table 8.1 (Continued) Common types of instruments for assessing performance of soil nailed structures

Parameters to be Monitored	Instruments	Details
		out unit. It provides a high resolution of horizontal profile of vertical deformation and settlement within the structure. The full profile of differential settlement can be obtained along the horizontal alignment.
	Extensometers	• Extensometers are another common type of instrument used to measure ground settlement. In general, a corrugated polyethylene pipe surrounded by rings of stainless steel wire at selected intervals is lowered into a borehole. A relatively rigid inner pipe, usually a PVC pipe, is coupled to the corrugated pipe prior to installation. A measurement device that can sense the presence of the steel rings is lowered and by comparing the depths of the signals with their corresponding initial depths, a settlement profile can be obtained. The main advantage of this instrument over settlement check point is that a settlement profile through the entire depth of strata can be obtained.
Forces Mobilized in Nail Reinforcement	Strain Gauges	• Soil nail reinforcement instrumented with strain gauges allows measurement of the load distribution during the construction stage, such as in the course of excavation and after the completion of the works. There are basically two common types of strain gauges, namely mechanical and electrical resistance types. Mechanical strain gauges are used to measure small changes of length between two reference points attached to a structural member, whereas electrical resistance strain gauges measure changes of resistance whenever there is strain. The two types of strain gauges can further be divided into surface mounted and embedment types. Among various types of strain gauges, vibrating wire type is probably the most common type that is used in the United States and Hong Kong. With the advancement of technology, there is an increasing trend of using optical grating and optical fiber types of strain gauges.

(Continued)

Table 8.1 (Continued) Common types of instruments for assessing performance of soil nailed structures

Parameters to be Monitored	Instruments	Details
Pressures Mobilized on Nail Heads or Facing	Load Cells	• Load cells are commonly installed at soil nail heads to provide information on the spatial and temporal distribution of actual loads that are developed at the nail heads and facing. Nonetheless, practical experience in the United States and Hong Kong suggests that load cells may occasionally provide inaccurate readings and the results should be interpreted with care. • Various types of load cells are available in the market, including telltale, mechanical, hydraulic, and electrical resistance load cells. Among the various types, electrical resistance load cells are probably the most commonly used for geotechnical engineering purposes. This type of load cells typically consists of Wheatstone-resistance bridges attached to the object to be measured with a known modulus. The resistance of the load cell is measured and is then correlated with the strain developed in the object and hence the load could be obtained.
Deformation	Survey Points	• The deformation of the concerned structures including the soil nailed structure may be measured directly by optical surveying methods or indirectly with electronic distance measuring (EDM) equipment. Reflector prisms are attached to the facing of a structure or installed on stable bases on the ground surface such that measurement could be made by these surveying methods. Optical or laser survey systems are typically capable of measuring horizontal and vertical displacements to an accuracy of 0.3 mm or better within a 100 m range. The ground deformation behind the soil nailed structure may also be assessed by monitoring an array or pattern of ground surface points established behind the structure and extending for a horizontal distance at least equal to the height of the structure. • In recent years, automated total stations have been used more frequently for optical surveying of retaining structures during construction. One advantage of an automated total station system is a reduction in reading errors by using an instrument that is semi-permanently installed. Also, the higher reading

(Continued)

Table 8.1 (Continued) Common types of instruments for assessing performance of soil nailed structures

Parameters to be Monitored	Instruments	Details
		frequency that can be accomplished with such systems permits discerning periodic, environment-associated movements from gravity-induced wall and ground movements. In projects of long duration, automated systems may result in savings as compared with traditional surveys.
Pore Water Pressure	Piezometers	• The information of pore water pressures is essential to understand the loading conditions of soil nailed structures. Piezometers are the most suitable instrument to suit the purpose. A piezometer is a device generally used to measure pore water pressure within the ground at selected depths. There is a variety of piezometers available in the market, which include open-hydraulic, closed-hydraulic, electrical (vibrating-wire), and pneumatic piezometers. Different types of piezometers have different merits and limitations. For example, pneumatic and electrical piezometers generally give rapid response time, whereas open piezometers are considered to be relatively simple, robust, and have greater long-term reliability. One should identify a suitable type of piezometer based on the defined objectives such as reliability, cost, availability, and response time.
Rainfall	Rain Gauges	• There are many types of rain gauges. The most common type is the ordinary rain gauge which is read manually using a measuring cylinder. Another common type is the autographic gauge which can be either of the tilting-siphon type or the tipping-bucket type. The recording chart of an autographic rain gauge is mounted on a drum which is driven by clockwork and typically rotates round a vertical axis once per day. Indeed, there has been significant technological advancement in rain gauges since the late 1970s with the use of the microprocessor. For example, a network of automatic rain gauges is being used in Hong Kong with rainfall readings available at one-minute intervals.

If any of the key parameters being monitored, such as ground movements and groundwater levels, approach or exceed the threshold values as defined at the design stage, appropriate mitigation measures or a revised construction sequence will be implemented.

Under normal circumstances, soil nails installed using the "drill-and-grout" method should not have any significant adverse water-damming effect on the hydrogeological regime. However, in cases of excessive grout leakage, the cement grout could reduce significantly the permeability of the ground, and groundwater may be dammed up. If there is a concern about the occurrence of the water-damming effect, piezometers should be installed immediately behind the anticipated extent of the soil nailed zone such that monitoring can be undertaken prior to and after soil nailing works to ascertain the effect. In planning the locations of piezometers, their potential of being blocked by the leaked grout should also be considered.

Although post-construction or long-term monitoring of soil nailed structures is generally not required, it may be carried out occasionally under special circumstances. For example, if a permanent nailed wall is constructed in the United States that involves critical or unusual construction (e.g. walls greater than 10 m high, widening under existing bridges, walls with high external surcharge loading, etc.), performance monitoring will be carried out for a period of at least 2 years after construction. The instrumentation for such walls usually includes inclinometers, survey points, load cells, and strain gauges. This enables monitoring of structural deformation and stress development in the nails and wall facing as a function of time and environmental changes, such as winter freeze-thaw cycle. In France, if the height of a nailed wall is greater than 10 m, or the structures to be supported by the wall are sensitive to movements or a sloping site is involved, monitoring of the wall throughout its service life is required. In this case, monitoring of ground movements, tensile forces in the nails and groundwater levels will be carried out as necessary. In Hong Kong, if an earth structure is reinforced by permanent soil nails that carry sustained loads, monitoring of the ground movements and loads mobilized along representative soil nails will be carried out during the construction stage and for an extended period after completion of works (normally at least two wet seasons after construction). Inclinometers are normally installed to obtain the full vertical profile of the horizontal ground movement. Monitoring of piezometric pressures is also carried out to aid the interpretation of deformation data.

Apart from ground movements and loading conditions of soil nails, durability is another factor that can govern the long-term performance of a nailed structure. If the durability of soil nails is of grave concern, e.g. nails installed in aggressive ground and the envisaged consequence of failure of the soil nailed structure would be catastrophic, it may be necessary to install sacrificial nails adjacent to the structure where these nails and the associated soil can be tested at intervals throughout the service life of the structure. For example, for medium- or long-term soil nailed walls in France, sacrificial short nail bars (usually 1 m to 1.5 m long) will be installed in conjunction with the wall construction. The bar samples will then be exhumed in regular intervals during the service life of the wall for inspection and testing where necessary. In Hong Kong, as part of the research program on durability of soil nails, sacrificial nails were installed at several sites of various ground aggressivity in the early 2000s. These nails were protected by different corrosion resistant measures, including hot-dip galvanizing, epoxy coating, stainless steel cladding, and solid stainless steel. Plain steel soil nails without corrosion protection measures were also installed as control specimens. The nails were then exhumed and tested at regular time intervals starting from 2010, so as to gain more knowledge about the long-term performance of different corrosion measures under different corrosive environments. More details about the study can be found in Section 5.3 of Chapter 5.

8.2.4 Case studies

8.2.4.1 40 m high open excavation in Brazil

Sayao et al. (2005) and Nunes et al. (2009) presented a case in which a 40 m high and 50 m long cut slope was formed by excavation. The slope was reinforced by soil nails and was instrumented to understand the fundamental behavior of the excavation and the nails during and after construction. The site is situated in Niteroi, a coastal city of Brazil, with the geology comprising heterogeneous gneissic residual soils. The upper 10 m layer is basically mature residual soil, predominantly red sandy clay, whereas the lower stratum is young residual soil of clayey sand with a lighter color. A 40 m high slope, with four batters and an angle ranging from 45° to sub-vertical, was formed by excavation at 2 m intervals progressively in gneissic residual soil. The excavation started in March 2004 and was completed in August of the same year. The upper two batters of about 20 m high comprised an unreinforced cut with grass as the surface cover. The lower two batters, on the other hand, were reinforced by soil nails made of 25 mm diameter steel bars with length varying from 15 m to 24 m. The soil nails were installed progressively with the excavation by the "drill-and-grout" method at a spacing of 2 m both vertically and horizontally. Figure 8.1 shows a typical cross-section of the soil nailed excavation with the installed instruments.

The instrumentation program focused on understanding the behavior of the excavation in respect of ground deformation and nail force mobilization. The instruments included inclinometers, tell-tales, and strain gauges. Inclinometers and tell-tales were employed to monitor the ground deformation. Two inclinometers, I_1 and I_2, were installed vertically at two boreholes of 50 mm in diameter to a depth of 16.7 m and 30 m, respectively (see Figure 8.1). The two inclinometers were situated at two different elevations and were both about 1 m away from the excavation face. It was intended

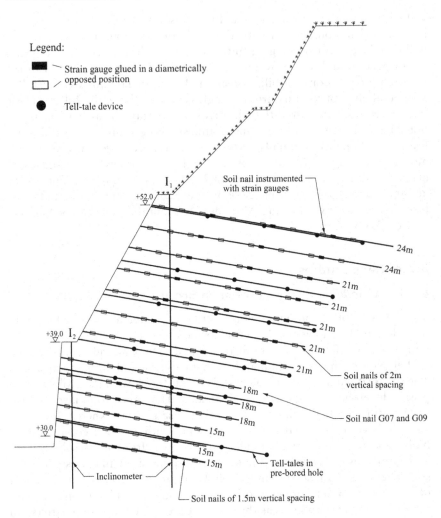

Legend:

▬ ⌐ Strain gauge glued in a diametrically
▢ ⟋ opposed position

● Tell-tale device

Soil nail instrumented
with strain gauges

24m
24m
21m
21m
21m
21m
21m

Soil nails of 2m
vertical spacing

18m
18m
18m

Soil nail G07 and G09

15m
15m
15m

Tell-tales in
pre-bored hole

Inclinometer

Soil nails of 1.5m vertical spacing

Figure 8.1 Cross-section of a typical instrumented nailed excavation. (Modified Nunes et al., 2009.)

to locate the zone of large shear strains within the reinforced soil mass and monitor the rate of deformation during and after excavation. The groundwater level was found to be below the bottom of the boreholes. In addition to inclinometers, six layers of tell-tale devices were installed at six predrilled holes at selected elevations to detect the potential failure zone (solid black lines with dots in Figure 8.1). The devices were positioned in the predrilled holes of 100 mm in diameter at 10° inclination below horizontal. For each pre-drilled hole, four tell-tales were provided at 1.5 m, 7 m, 14 m, and 28 m, respectively, from the excavation face. Each tell-tale device was connected by a cable to a dead weight of about 0.5 kg at the cut surface with pulley, for keeping the cable stretched. Once ground deformation occurred, displacement readings could be indicated on millimeter scale by a needle attached to the cable. In regard to mobilization of tensile forces, five rows of nails were instrumented in two central vertical sections of the cut. Strain gauges (electrical extensometers) were glued onto ten steel bars to monitor the distribution of tensile forces. Moreover, some strain gauges were also glued in a diametrically opposed position, so as to measure the bending moment induced along the soil nail.

The 40 m excavation was completed in about 6 months and the monitoring went on for 2 years after construction was completed. In general, progressive increase in horizontal displacement was registered as the excavation progressed as indicated in Figure 8.2. It is interesting to note that the horizontal ground deformation continued to occur after the completion of excavation in August 2004. According to Lima (2007), the continuing ground deformations were probably caused by excavations in a neighboring construction site. Sayao et al. (2005) pointed out from the records of strain gauges that the distribution of tension along a nail bar was non-uniform. The nail force started with a positive value near the excavation face, increased and reached a maximum near the middle of the bar, and decreased to zero at the far end of the nail. As the excavation was covered by shotcrete facing, the tensile nail force as mobilized near the surface could be due to the fixed end condition. Figure 8.3 shows a typical temporal variation of tensile forces registered at two strain gauges on nails G07 and G09, respectively, situated at 3 m from the excavation face. One may notice that there was a progressive increase in the mobilized tensions during excavation works, and there was a tendency of load steadiness after completion of construction works.

8.2.4.2 50 m high open excavation in Hong Kong

Kwong & Chim (2015) presented a case history in Hong Kong wherein a 50 m high steep slope was formed by open excavation to facilitate the construction of a railway box station (Figure 8.4). The site geology comprises a thick layer of fill material overlying decomposed granite and granitic bedrock. The works involved cutting into a mantle of fill up to 30 m thick and some 20 m into the underlying completely and highly decomposed granite. The cutting and strengthening works were carried out in stages as excavation proceeded downward until the design formation level was reached. As indicated in Figure 8.5, the upper batter was a 30 m high and 45° steep excavation in fill, whereas the lower batter was a 20 m high and 75° cut in decomposed granite. The excavation was reinforced by some 700 glass fiber reinforced polymer (GFRP) nails, with a diameter of 40 mm and lengths varying from 16 m to 24 m. The nails were installed at an inclination of 20° below horizontal with a vertical spacing of 2 m and 1.5 m in fill and decomposed granite, respectively. The horizontal spacing of all the nails was 2 m.

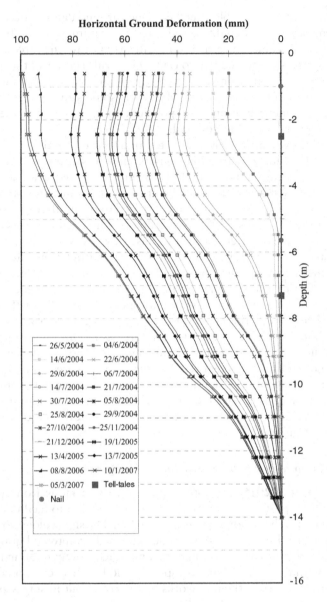

Figure 8.2 Horizontal ground deformation profile at inclinometer I2. (Modified Nunes et al., 2009.)

As very steep and high slopes were formed as part of the site formation works, an instrumentation and monitoring plan was in place to validate the performance of the soil nailed structure during the course of construction. Various types of instruments were selected to suit the monitoring purposes, which included 20 inclinometers, numerous ground settlement markers and

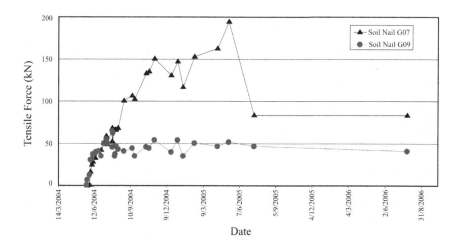

Figure 8.3 Tensile forces registered in strain gauges at 3 m from slope face at nail G07 and G09. (Modified Nunes et al., 2009.)

piezometers spreading over the whole site, numerous surface prisms on the shotcrete surface, and utility monitoring points. In addition, about 5% (i.e. 35 nos.) of soil nails were strain gauged and also had surface prisms installed at their respective nail heads (Figure 8.5). As part of the monitoring plan, the response corresponding to "alert," "action," and "alarm" (AAA) levels based on empirical rule of thumb, past experience, and structural assessment of the anticipated ground deformation effects on the concerned facilities in the vicinity of the site was developed as shown in Table 8.2.

The excavation and nail installation started in March 2012. As the construction works proceeded, horizontal deformation and vertical settlement started to increase. When about a 10 m high slope was formed in May 2012, the following monitoring readings were recorded:

- *Inclinometers*: 40 mm and 16 mm horizontal movements were recorded at the middle and crest of the slope (c.f. Action Level is 40 mm).
- *Settlement markers*: 30 to 35 mm vertical settlements were recorded at the slope crest (c.f. Alert Level is 25 mm).
- *Prisms*: 10 to 20 mm horizontal movement, and 5 to 15 mm vertical movement were recorded.
- *Strain gauges*: 60 to 100 kN tensile forces were developed in the soil nails.
- *Piezometers*: groundwater table was below the excavation level.

Ground deformation analysis using the finite element method was carried out at the design stage and the predicted horizontal movement at that particular excavation level was about 20 mm. The recorded horizontal movement was already 40 mm which had doubled the prediction. Also, there was

Figure 8.4 General setting of the excavation site. (Modified Kwong & Chim, 2015.)

a grave concern from other monitoring records as shown above that if the excavation works were to continue to the final formation level, and with the prevailing rate of increase in horizontal movement and settlement, excessive movements could cause collapse if no action was taken. As a result, remedial works including the changing nail installation technique from unsupported drillhole to drilling with temporary casing, lengthening the nails at some critical locations, installation of additional nails and raking drains, paving the entire slope crest with concrete, etc., were proposed. As the result of these remedial measures and with close monitoring, further increase in the ground deformation became steady and was under control. The excavation continued down to the final formation level and the final maximum horizontal movement and vertical settlement were recorded to be 70 mm and 90 mm, respectively. No distress was observed on the slope as well as the facilities in the vicinity of the site.

8.3 MAINTENANCE

8.3.1 General

In principle, a soil nailed structure should be designed to withstand any combination of the anticipated adverse effects of all the deteriorating

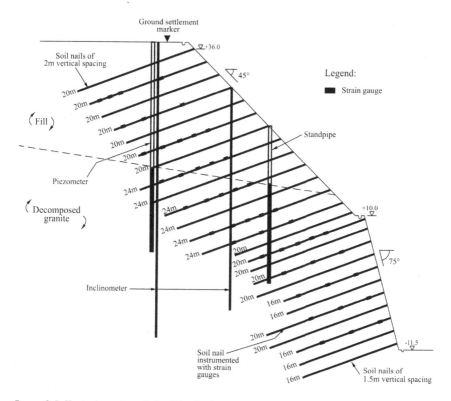

Figure 8.5 Typical section of the 50 m high steep slope and monitoring points. (Modified Kwong & Chim, 2015.)

actions that the structure would be exposed to during its service life. It is also desirable that soil nailed structures should require as little maintenance as possible. Nonetheless, it is good practice that regular maintenance inspections and the necessary associated works are carried out to keep a soil nailed structure in good condition and ensure that it can perform its intended function properly throughout its service life. The inspections and the maintenance works can reduce the probability of failure of the soil nailed structure.

8.3.2 Maintenance inspections and the associated works

Maintenance inspections and the associated works should be carried out at regular intervals throughout the service life to make sure that the soil nailed structure can perform its intended function properly and its safety will not be jeopardized. In Hong Kong, maintenance inspections are classified into two categories, namely routine maintenance inspection and engineer

Table 8.2 Response levels of the open excavation

Monitoring Parameters	Alert Level	Action Level	Alarm Level
Ground Settlement	25 mm	40 mm	50 mm
Ground Horizontal Movement	25 mm	40 mm	50 mm
Building Settlement/ Tilting	12.5 mm 1:600	20 mm 1:375	25 mm 1:300
Utilities Settlement/ Tilting	10 mm 1:600	16 mm 1:375	20 mm 1:300
Groundwater Drawdown	0.5 m below reference level	0.8 m below reference level	1 m below reference level

Note: Alert Level is set at 50% of maximum allowable level.

Action Level is set at 80% of maximum allowable level.

Alarm Level is set at 100% of maximum allowable level.

inspection. The scope, frequency, and personnel required to carry out each type of inspection are summarized in Table 8.3 (GEO, 2018).

In the UK, maintenance inspection programs are drawn up, depending on the risk category of the soil nailed structure (O'Donovan et al., 2020). Facilities for maintenance, especially for safe and efficient access for maintenance, should be incorporated as part of the design. Regular maintenance inspections and condition appraisal are recommended to check the following:

- Movement of facing and localized bulging.
- Movement of the soil nailed structure and existence of tension cracks at the crest.
- Condition of the drainage system.
- Condition of the soil nails and the facing, and the amount of corrosion damage.
- Condition of vegetation growth on the soil nailed structure, in particular if trees have grown too large that they might be blown over and cause damage.
- Degradation of the soil nailed structure caused by animal or human activity.

O'Donovan et al. (2020) recommended that the frequency of inspections should be no greater then five-year intervals, at least for the first 10 years upon completion of the soil nailing works. After that, the inspection frequency should be determined based on knowledge of the earth structure.

8.3.3 Maintenance manual

It is good practice that the engineer who designs a soil nailed structure would prepare a maintenance manual as part of the design services. A maintenance manual helps the owners or parties responsible for maintaining the structure

Table 8.3 Types of maintenance inspections in Hong Kong (GEO, 2018)

Types of Maintenance Inspections	Scope	Frequency	Personnel Requirements
Routine	• Clearance of accumulated debris from drainage channels and surface of the structures. • Repair of cracked or damaged surface drainage channels. • Repair or replacement of cracked or damaged facings. • Unblocking of weep holes and outlet drain pipes. • Removal of any vegetation that has caused severe cracking of facings and surface drainage channels. • Trim overgrown vegetation. • Replant vegetation where it has died. • Repair or replacement of rusted steel furniture. • Maintenance of other hard landscape items.	Once every 1 to 2 years (depends on the risk level of the soil nailed structure)	Any responsible person with no professional geotechnical knowledge
Engineer	• To identify all visible changes and signs of distress, in particular changes since the previous stability assessment if this has been carried out, and any discrepancies between records and site conditions, which could have implications on stability of the soil nailed structure, and to judge whether these might be significant. • To check that routine maintenance inspections have been carried out and documented satisfactorily. • To assess the adequacy of routine maintenance works and supplement the list of engineering or landscaping items requiring routine maintenance if necessary. • To advise whether a stability assessment of the soil nailed structure is necessary. • To recommend the necessary maintenance works.	Once every 5 to 10 years (depends on the risk level of the soil nailed structure)	A professionally qualified geotechnical engineer

to appreciate the maintenance requirements. A maintenance manual for a soil nailed structure normally includes the following information:

- A plan showing the soil nailed structure to be maintained.
- Records containing general information on the soil nailed structure, including record photographs.
- As-built plans and typical cross-sections of the soil nailed structure, including details of facing, surface drainage, subsurface drainage, access points, and details of soil nails.
- Layout plans of utilities and services on or adjacent to the soil nailed structure, and proper documentation of any special features (e.g. ducting system) related to the services.
- A list of engineering and landscape items requiring routine maintenance.
- Recommendations on the frequency of routine maintenance inspections, engineer inspections for maintenance, and regular checks of buried water-carrying services (including ducting systems) on or adjacent to the soil nailed structure where appropriate.
- Outline of the basis of design and/or findings of stability assessments, including the risk level of the soil nailed structure.

8.4 CONCLUDING REMARKS

Monitoring and regular maintenance play an important role in upkeeping the function of soil nailed structures throughout the process of construction and/or service of the structures. They can also enrich our understanding and knowledge about the short-term and long-term performance of soil nailed structures, so as to enhance the design and construction process of similar structures in future. A monitoring plan and maintenance manual should be developed as part of the design and construction process. In this chapter, the key issues involving monitoring and maintenance are identified and discussed.

Monitoring may be carried out for different reasons. These include verification of design assumptions, provision of information as part of the observational method during the construction stage for construction control, and collection of information for research purposes. In practice, depending on the purpose of the monitoring, a variety of parameters could be monitored.

The parameters commonly specified in a monitoring plan include ground deformation, magnitude and distribution of loads developed in the soil nails, nail heads, and slope/excavation facings, groundwater condition, soil aggressivity, temperature, and rainfall. It is essential for a designer to define the objectives of monitoring and identify the suitable instrumentation. The instruments that are commonly employed in monitoring a soil nailed structure include inclinometers, settlement markers, survey points, strain gauges, piezometers, rain gauges, etc. Readers should familiarize themselves with the merits and limitations of these instruments in order to select the suitable ones for effective implementation of monitoring plan.

Although a soil nailed structure generally requires little maintenance, a poorly maintained structure could deteriorate and become so vulnerable that it may collapse and cause casualties or damage to property. Regular maintenance is considered the essential and effective means to upkeep the condition and function of a soil nailed structure in the long term. The maintenance practices adopted in different places have been presented. It is important to develop a suitable maintenance manual as part of the design and construction process such that the critical elements can be identified and properly maintained throughout the service life of the structure.

REFERENCES

Byrne, R.J., Cotton, D., Porterfield, J., Wolschlag, C. & Ueblacker, G. (1998). Manual for Design and Construction Monitoring of Soil Nail Walls. Federal Highway Administration, US Department of Transportation, Washington, D.C., USA, Report No. FHWA-SA-96-069R, 530 p.

Clouterre (1991). French National Research Project Clouterre: Recommendations Clouterre (English Translation 1993). Federal Highway Administration, US Department of Transportation, Washington, D.C., USA, Report No. FHWA-SA-93-026, 321 p.

Dunnicliff, J. (1993). *Geotechnical Instruments for Monitoring Field Performance*. Wiley, New York, 600 p.

GEO (Geotechnical Engineering Office) (2008). *Guide to Soil Nail Design and Construction (Geoguide 7)*. Geotechnical Engineering Office, Civil Engineering and Development Department, HKSAR Government, Hong Kong, 97 p.

GEO (2018). *Guide to Slope Maintenance (Geoguide 5)*. Geotechnical Engineering Office, Civil Engineering and Development Department, HKSAR Government, Hong Kong, 116 p.

Kwong, A.K.L. & Chim, J.S.S. (2015). Performance monitoring of a 50m high 75° cut slope reinforced with soil nail bars made from glass fiber reinforced polymer (GFRP) at Ho Man Tin Station. In Proceedings of the HKIE Geotechnical Division 35th Annual Seminar, the Hong Kong Institution of Engineers, Hong Kong, pp. 141–149.

Lazarte, C.A., Elias, V., Espinoza, R. D., & Sabatini, P.J. (2003). Geotechnical Engineering Circular No. 7: Soil Nail Walls. Federal Highway Administration, US Department of Transportation, Washington, D.C., USA, Report No. FHWA0-IF-03-017, 182 p.

Lazarte, C.A., Robinson, H., Gomez, J.E., Baxter, A., Cadden, A. & Berg, R. (2015). Geotechnical Engineering Circular No. 7: Soil Nail Walls: Reference Manual. Federal Highway Administration, US Department of Transportation, Washington, D.C., USA, Report No. FHWA-NHI-14-007, 385 p.

Lima, A.P. (2007). Behaviour of a nailed excavation in gneissic residual soil (in Portuguese). PhD thesis, Civil Engineering Department, Pontifical Catholic University of Rio de Janeiro—PUC Rio, Brazil, 431p.

Nunes, A.L.L.S, Sayao, A.S.F.J. & Lima, A.P. (2009). Behaviour of nails reinforcing an open excavation in gneissic residual soil. In Proceedings of the 17th International Conference on Soil Mechanics and Geotechnical Engineering, pp. 1939–1942.

O'Donovan, J., Feudale Foti, E., Turner, M., Irvin, C. & Mothersille, D. (2020). Grouted Anchors and Soil Nails - Inspection, Condition Assessment and Remediation. Construction Industry Research & Information Association, London, UK, CIRIA Report No. C794, 193 p.

Sayao, A.S.F.J., Lima, A.P., Springer, F.O., Nunes, A.L.L.S., Dias, P.H.V. & Gerscovich, D.M.S. (2005). Design and instrumentation aspects of a 40m high nailed slope. In Proceedings of the International Conference on Soil Mechanics and Geotechnical Engineering. Osaka, Japan.

Slope Indicator (2004). Guide to Geotechnical Instrumentation. Slope Indicator Company, 50 p.

Transport for NSW (2016). Geotechnical Instrumentation and Monitoring Guidelines. Transport for NSW, NSW Government, New South Wales, Australia, 60 p.

Chapter 9

International standards of good practice

9.1 INTRODUCTION

A good standard of practice in design and construction of soil nails is essential to ensure good quality and satisfactory performance of a soil nailed structure throughout its serving period. Different countries have developed and adopted their own tailor-made design codes and construction specifications for carrying out soil nailing works. Although there may be some variations among these design standards or codes of practice, the underlining principle of soil nailing technique is the same, i.e. to improve the stability of slopes, retaining walls, embankments, or excavations by effectively mobilizing tensile forces in the soil nails. In general, both the Allowable Stress Design (ASD), also known as the Working Stress Design (WSD), and the Limit State Design (LSD), also known as the Load and Resistance Factor Design (LRFD), are the two common design approaches being adopted in various standards and codes of practice. All the design codes require consideration of potential external and internal failure modes in assessing the overall stability of a soil nailed structure. By applying appropriate sets of safety factors (global or partial) against different potential failure modes as recommended by the design standards, the uncertainties associated with the design and construction process would be addressed. Apart from design aspects, good specifications for constructing and maintaining the soil nails are essential to ensure good quality of works, and safe and proper functioning of a soil nailed structure. The specifications should provide guidance on or requirements of all the critical activities and issues during the construction stage as well as throughout the service period of the soil nailed structure. As a minimum, they should cover issues associated with soil and groundwater conditions, quality of construction materials, installation methods, qualification requirements for contractors and supervision personnel, level of site supervision, testing requirements, and items for maintenance or monitoring.

In this chapter, selected standards of practice or guidance on the design and construction of soil nails developed and promulgated in various places, including France, the UK, the United States, the Nordic countries, Japan, and Hong Kong, are presented and compared. It is intended to give the readers an overall picture of the state-of-the-art standards of good practice in soil nailing. The selected codes of practice or guidelines covered in the review are summarized in Table 9.1.

9.2 FRENCH PRACTICE

A large research project, French National Research Project, Clouterre, was carried out between 1986 and 1989 in France with the objective to study the behavior of soil nailed retaining walls during construction, in service, and at failure (see Section 1.3.2 in Chapter 1 for more details about the research project). The findings of the research project were consolidated into the publication of the first systematic and comprehensive code of practice by the Presses de l'École Nationale des Ponts et Chaussées in 1991 (English version in 1993) for application of soil nailing in France, namely Recommendations Clouterre 1991 (Clouterre, 1991). To date, the code is still used in France, which promulgates state-of-good-practice recommendations for application of soil nailing in earth structures, particularly for reinforcing retaining walls and excavations. The guidance covers investigation, design, and construction of soil nails for short-term (service life less than 18 months), medium-term (18 months to 30 years), and long-term (30 to 100 years) applications. Advice on maintenance and monitoring is, however, comparatively limited.

The code advocates using the Limit State Design (LSD) approach to design soil nailed structures. A set of partial factors to cater for different design scenarios is recommended. It requires identification and examination of both the external and internal failure modes in assessing the overall stability of a soil nailed structure. Application of limit equilibrium methods in carrying out stability assessment is recommended throughout the publication. Coupling with limit equilibrium methods, the code recommends the use of appropriate analytical methods, such as the method of slices (e.g. Fellenius or Bishop's method), for assessing external stability of a nailed structure. Guidance on deformation analysis is also given to facilitate such assessment when required under special circumstances. Recognizing the difficulty in carrying out numerical analysis, the code suggests adopting a pragmatic approach to estimate wall displacements by using empirical correlation between deformation and pertinent parameters, such as wall height, spacing and length of soil nails, and soil characteristics. For internal stability, the code suggests examination of various potential failure modes including pullout and facing failures. However, unlike most of the other standards of good practice promulgated in different places, the French code requires consideration of the failure mode of nail reinforcement under combined actions of tension, shear, and bending. For pullout failure mode, the code requires only the bond strength at the interface between grout and the ground to be assessed. In the estimation of pullout resistance, preliminary

Table 9.1 Summary of standards of good practice of soil nailing in different places

Country/ Place	Standard of Good Practice	Organization
France	French National Research Project Clouterre—Recommendations Clouterre (Clouterre, 1991)	Presses de l'École Nationale des Ponts et Chaussées, France (English Translation by the US Federal Highway Administration)
UK	Design Methods for the Reinforcement of Highway Slopes by Reinforced Soil and Soil Nailing Techniques (DOT, 1994)	The Department of Transport
	Soil Nailing—Best Practice Guidance (CIRIA Report No. C637) (Phear et al., 2005)	Construction Industry Research & Information Association, London
	Execution of Special Geotechnical Works—Soil Nailing (BS EN 14490:2010) (BSI, 2010)	British Standards Institution, London
	Code of Practice for Strengthened/ Reinforced Soils—Part 2: Soil Nail Design (BS 8006-2:2011) (BSI, 2013)	British Standards Institution, London
	O'Donovan, J., Feudale Foti, E., Turner, M., Irvin, C. & Mothersille, D. (2020). Grouted Anchors and Soil Nails - Inspection, Condition Assessment and Remediation (CIRIA Report No. C794) (O'Donovan et al., 2020)	Construction Industry Research & Information Association, London
United States	Manual for Design and Construction Monitoring of Soil Nail Walls (Byrne et al. 1998)	Federal Highway Administration, US Department of Transportation, Washington, D.C.
	Geotechnical Engineering Circular No. 7: Soil Nail Walls – Reference Manual (Lazarte et al. 2015)	Federal Highway Administration, US Department of Transportation, Washington, D.C.
Nordic Countries	Nordic guidelines for reinforced soils and fills (Rogbeck et al. 2005)	Nordic Geosynthetic Group
Japan	Design and Construction Guidelines for Soil Nailing (JHPC, 1998)	Japan Highway Public Corporation
Hong Kong	Geoguide 7: Guide to Soil Nail Design and Construction (GEO, 2008)	Geotechnical Engineering Office
	GEO Publication No. 1/2009: Prescriptive Measures for Man-made Slopes and Retaining Walls (GEO, 2009)	Geotechnical Engineering Office

design values are provided based on empirical correlation with pertinent parameters, such as soil type and pressuremeter test data. Guidelines are also given on the estimation of pullout resistance by means of field pullout tests (see Chapter 6 for more details).

Regarding the layout and dimensions of soil nails, guidance is given on nail density, spacing, length, and inclination. As a rule of thumb, a regular nail pattern is recommended, and the average length of nails depends on the installation method and the height of the soil nailed structure. For driven nails, the length should normally be about 0.5 to 0.7 H (where H is the height of the soil nailed structure), and the density of nails should be about 1 to 2 nails per m² (i.e. spacing of about 0.7 m to 1 m). The length should, however, be 0.8 to 1.2 H if "drill-and-grout" nails are used, and the recommended density becomes 0.15 to 0.4 nails per m² (i.e. spacing of about 1.5 m to 2.5 m). In general, the minimum nail density should be kept to at least one nail per 6 m². Regarding nail inclination, the document suggests an angle of 5° to 15° below horizontal to be adopted. Some design charts are also provided to determine the preliminary design of soil nail walls, and empirical correlations are provided to estimate the maximum horizontal and vertical displacements at the top of a soil nailed wall. In designing the appropriate corrosion protection measures to nail reinforcement, the code recommends carrying out a comprehensive ground corrosivity assessment. Depending on the results of the assessment, guidance is provided for the selection of corrosion protection measures, which include provision of sacrificial steel thickness, coating to nail reinforcement (galvanizing or non-metallic), and plastic encapsulation, to suit different levels of ground corrosivity (see Chapter 5 for more details).

Apart from design aspects, the code also provides recommendations on various aspects in relation to construction and quality control of soil nailing works. These include selection of the nail installation method and equipment, construction sequence, construction problems that are commonly encountered and the respective suggested preventive and mitigation measures, site testing, and quality control, etc. Sample specifications and inspection focus are also provided by the document to facilitate the execution of soil nail construction works.

9.3 UK PRACTICE

9.3.1 Design Methods for the Reinforcement of Highway Slopes by Reinforced Soil and Soil Nailing Techniques (HA 68/94)

The first standard specifically for design of soil nails in the UK is in the form of an advice note, entitled HA 68/94—Design Methods for the Reinforcement of Highway Slopes by Reinforced Soil and Soil Nailing Techniques, which was published by the Department of Transport in 1994 (DOT, 1994). The advice note provides guidance on the design requirements for the strengthening of highway earthworks using reinforced soil and soil nailing techniques. The design guidelines are basically applicable for reinforcing new earthworks, widening of existing embankments, steepening of existing cuttings, and repairing of failed slopes. The nominal design life of the reinforced earth structures is 60 years. The LSD approach incorporating partial factors is adopted for the design against both the ultimate and serviceability limit states. A set of partial factors each for the nominal driving and resisting forces is recommended. The nominal driving forces include self-weight of soil, surcharge and pore water pressures, and the resisting forces are essentially the shear strength of soil and the reinforcement tensile force. The resistance contributed by the bending stiffness of the reinforcement is ignored. The ultimate limit state corresponds to the formation of a collapse mechanism in the nailed structure (i.e. the upper bound solution), whereas the serviceability limit state focuses on ground deformation that may affect the function of the structure or adjacent facilities or services. Examination of both the external and internal stability is required in assessing the overall stability of a soil nailed structure. Moreover, it is required to assess the effects of expansion of the soil due to swelling or freezing although no specific guidelines have been provided.

A limit equilibrium approach based on a two-part wedge method is recommended for assessing the external stability. The advice note considers it is preferable to adopt this approach because it provides a simple method for obtaining safe and economical solutions. It is inherently conservative when compared to more exact solutions and allows simple hand check calculations to be carried out. For internal stability, the standard requires consideration of the tensile failure of reinforcement, and bond failure between grout and soil as well as grout and reinforcement. Bending or shear failure of nail reinforcement, however, can be disregarded. In comparison with stability assessment, limited guidance is given on deformation analysis.

The advice note provides some guidance on the layout and dimensions of soil nails. For example, the maximum vertical spacing of nails is limited to 2 m, whereas that for the horizontal spacing should not exceed the maximum value of vertical spacing. In theory, there are many possible combinations of nail spacing and length which satisfy the stability requirement. To facilitate the designer in this respect, an empirical formula for optimizing the vertical spacing of soil nails is provided by varying the nail spacing and length throughout the structure, so as to prevent local over-stressing of any layer of reinforcement. In addition, the inclination of soil nails is recommended to be 10° below horizontal and it is suggested advantageous to insert the first layer of nails at a steeper angle than the others in order to increase the pullout resistance. In estimating the pullout resistance of soil nails, the advice note recommends use of the effective stress approach coupled with verification by the field pullout test. As an alternative, the standard also suggests carrying out the field pullout test in order to obtain a more realistic design value of site-specific pullout resistance.

In the context of durability, the advice note recommends carrying out a soil corrosivity assessment in order to classify the corrosion potential of the site. Depending on the outcome of the assessment, various corrosion protection measures could be selected to suit the required protection level. These measures include sacrificial steel thickness, galvanizing or other protective coating, grout cover, and corrugated plastic sheathing within grout annulus. As the scope of the document is limited to design aspects, no guidance is given on construction, monitoring, and maintenance of soil nailed structures.

9.3.2 Code of Practice for Strengthened/Reinforced Soils, Part 2: Soil Nail Design (BS 8006-2: 2011)

Apart from advice note HA 68/94, another standard specifically for design of soil nails in the UK is BS 8006:1995—Code of Practice for Strengthened/ Reinforced Soils and Other Fills, which was published by the British Standards Institution in 1995. Following the publication of Eurocodes for geotechnical design, in which Eurocode EN 1997-1:2004—Geotechnical Design explicitly excludes the design of reinforced soil structures and slopes, the British Standards Institution decided to carry out a review of BS 8006 and subsequently issued BS 8006-2: 2011 Code of Practice for Strengthened/Reinforced Soils, Part 2: Soil Nail Design in 2013 (BSI, 2013) to harmonize the design approach of soil nailing with other geotechnical structures designed using Eurocodes. This part of BS 8006 provides the latest technical guidance on designing reinforced earth structures, including slopes, embankments, and excavations, using soil nails. The guidelines provided by the standard are applicable to soil nailed structures with a service life ranging from a few months to 120 years. The LSD approach is adopted with a view to aligning the design philosophy with other geotechnical designs under the Eurocodes. Both ultimate limit state and serviceability limit state would be considered. The former is associated with instability, collapse, or failure of the nailed structures, whereas the latter corresponds to excessive deformation or when the serviceability of the structures is impaired. Partial factors are recommended to the actions, material properties, and soil nail resistance.

Similar to other standards and codes of practice, soil nails are designed as tensile elements without consideration of the shear and bending effects. Both the external and internal stability have to be considered in assessing the overall stability of a soil nailed structure. The global failure mode, including sliding, rotational, and deep-seated failure, should be analyzed in the external stability assessment. Although 2-D limit equilibrium methods coupled with the method of slices (e.g. Bishop's simplified method) or bi-linear slip methods (e.g. two-part wedge method) are recommended for assessing the external stability, the observational method, in which the design of the nailed structure is reviewed during construction, may also be used. The standard also recognizes the need for carrying out numerical analysis under special circumstances. For example, numerical analysis may be required when a clearer understanding of the soil-structure interaction or deformations in complex geological or geometric situations is needed. In selecting appropriate numerical methods, one should take into account the nature of ground conditions, interaction of the structural components with the ground, and compatibility of strains at the limit state being investigated. In addition, it would be preferable to calibrate the numerical or constitutive models with precedent experience, supplemented with construction monitoring. The standard, however, does not provide detailed advice on the essential requirements when undertaking numerical modeling. Apart from external stability, consideration of various potential internal failure modes particularly on the tensile failure of nail reinforcement, bond failure at the interface of the ground/grout and reinforcement/grout, punching failure of nail heads, and facing failure are required. Regarding estimation of pullout resistance of soil nails, various approaches are recommended. They include making reference to published data, empirical correlation with soil tests such as SPT N values and pressuremeter test results, effective stress method, total stress method, and field pullout test. Serviceability limit state mainly refers to excessive deformation and the standard suggests use of empirical means to assess the serviceability of the nailed structures.

The standard also gives advice on the design of layout and dimensions of soil nails. For example, a regular triangular nail pattern should be adopted. The spacing (vertical and horizontal) should generally range from 1.5 m to 3.0 m and 1.0 m to 2.0 m for gentle (less than 45°) and intermediate (45° to 60°) slopes, respectively. For steep slopes (greater than 60°), the vertical and horizontal spacing should fall within 0.75 m to 1.5 m and 0.5 m to 2.0 m, respectively. Furthermore, the standard suggests adoption of uniform vertical spacing or decreasing the vertical spacing with depth in order to avoid possible overstressing of nail locally. Regarding nail length, it should normally be about 0.5 to 2 H and 0.5 to 1.5 H (where H is the height of the soil nailed structure) for gentle and intermediate slopes, respectively. For steep slopes, the nail length becomes 0.5 to 1.2 H. No guidance is, however, given for nail inclination.

To enhance the long-term performance of soil nails, the standard requires carrying out an assessment of degradation risk, so as to classify the corrosion level of the site. Based on the findings of the assessment, appropriate corrosion protection measures could be selected to suit the environment. The eligible measures include selection of suitable material for nail reinforcement, such as stainless steel and fiber composites, and coating on the nail reinforcement like galvanizing and epoxy, and plastic encapsulation. Similar to that in advice note HA 68/94, limited guidance is given by the standard on matters related to construction, monitoring, and maintenance of soil nailed structures.

9.3.3 Execution of Special Geotechnical Works— Soil Nailing (BS EN 14490: 2010)

To supplement BS 8006-2: 2011, which mainly covers design aspects of soil nails, the British Standards Institution published another document governing the construction practice of soil nailing in the UK, namely BS EN 14490: 2010—Execution of Special Geotechnical Works—Soil Nailing (BSI, 2010). This standard provides a full coverage and establishes the general principles of construction, supervision, testing, and quality control requirements for execution and long-term monitoring of soil nailing works. In particular, the standard provides comprehensive guidance on the practice aspects of soil nailing, testing of soil nailed structures, and a model specification.

9.3.4 Soil Nailing—Best Practice Guidance (CIRIA Report No. C637)

Another important publication that summarizes the good practice for application of the soil nailing technique in the UK is a guidance document entitled Soil Nailing—Best Practice Guidance (CIRIA Report No. C637) (Phear et al., 2005). The document was published by the Construction Industry Research & Information Association in 2005, which promulgates state-of-the-practice about the design, construction, testing, and maintenance of soil nailed walls and slopes. The publication is based on literature review, consultation with experts and practitioners in the industry, existing standards, and case studies. Apart from promulgating the best practice at the time and providing a guide for routine use, the document encourages confident and more widespread use of the soil nailing technique in the UK.

The guide covers various aspects about the soil nailing technique, including the history of development, site investigation, risk-based conceptual and detailed design, durability, critical construction issues, quality control and testing, and maintenance issues. Indeed, the standards for application of the soil nailing technique being adopted in the UK currently, i.e. BSI (2010, 2013), are mostly developed on the basis of this guidance document.

9.3.5 GROUTED ANCHORS AND SOIL NAILS - INSPECTION, CONDITION ASSESSMENT AND REMEDIATION (CIRIA REPORT NO. C794)

The publications as described in Section 9.3.1 to 9.3.4 are designated for design and construction of soil nails with little or no guidance on their post-construction inspection, appraisal and remediation. In order to address the demand for these aspects from the industry, the Construction Industry Research & Information Association published a supplement document entitled Grouted

Anchors and Soil Nails - Inspection, Condition Assessment and Remediation (CIRIA Report No. C794) in 2020 (O'Donovan et al., 2020).

For the part relating soil nailing, the report provides guidance on the approach for inspection, condition assessment and remediation of soil nails, facing and drainage. The planning and details of inspection, condition assessment and load testing of soil nails are given. Some common issues of failed soil nailed structures and the corresponding remedial works in the form of case studies are presented in the report. They include bulging or failure of facings, insufficient or defective drainage, bond failure at resistant zone, etc.

9.4 US PRACTICE

9.4.1 Manual for Design and Construction Monitoring of Soil Nail Walls

The first comprehensive and well-structured design and construction standard for application of soil nailing in earth structures is the Manual for Design and Construction Monitoring of Soil Nail Walls published by the Federal Highway Administration of the US Department of Transportation in 1998 (Byrne et al., 1998). The manual specifically introduces the concept of soil nailing for highway applications and provides guidelines on site investigation, analysis, design, and construction of soil nailed walls. The guidelines are applicable to both permanent (75 to 100 years) and temporary (normally 18 to 36 months) applications. The focus of the manual is on the design aspects although selected construction related subjects, such as some technical issues during construction and preparation of specifications and contracts, are covered by the document.

Both the Allowable Stress Design (ASD) (also known as the Working Stress Design, WSD) and the Limit State Design (LSD) (also known as Load and Resistance Factor Design, LRFD, in the United States) approaches are introduced in the design procedures. However, as the information was insufficient to develop fully the LRFD guidelines at the time, the document adopts primarily the ASD approach for designing soil nailed walls. The manual recommends use of limit equilibrium analysis methods for stability analyses. It requires consideration of both potential external and internal failure modes in the course of assessing the stability of a soil nailed wall. Potential external failure modes include global, sliding, bearing, and overturning stability. In carrying out global stability analysis, use of limit equilibrium methods coupled with the slip circle or bi-linear slip method (e.g. two-part wedge method) is suggested. For internal stability, the manual recommends consideration of only two major failure modes, namely tensile failure of nail reinforcement and pullout failure at the interface between the grout and the ground. As the bond between grout and reinforcement is typically an order of magnitude higher, or more, than that between the ground and grout, the manual considers that it is not a critical failure mode to be considered in the design. Similarly, bending and shear capacity of nail reinforcement are not considered as critical, and hence they could be disregarded in the design. Similar to the practice of many other places, pullout resistance is considered as a key element in a soil nail design. The manual recommends a preliminary design value can be estimated based on local experience or empirical correlation with suitable soil tests such as SPT, followed by verification using the field pullout test. Alternatively, the design value can be obtained directly by carrying out field pullout tests during the site investigation stage.

Apart from stability, the manual requires consideration of serviceability requirements, such as control of wall deformation and surface cracking of wall surface. The manual advocates use of empirical approach to handle serviceability assessment. For example, empirical correlation between the maximum wall deformation and pertinent parameters, such as soil type and wall geometry compiled by the French code (Clouterre, 1991) from past experience, is presented as general guidance. For the checking of crack width, relevant structural codes have been promulgated in the United States. Design guidance on some special circumstances, such as frost protection, handling of seismic loads, and expansive soils, is also provided by the document.

In the design of nail layout and dimensions, the adoption of a regular grid pattern, including staggered arrangement, is recommended. The manual suggests that it is desirable to install nails close to the horizontal, if possible, within a range from 5° to 15° below horizontal. The nail length should normally be limited to 0.6 to 1.0 H (where H is the height of the wall). Also, the nail spacing should generally range from 1 m to 2 m, both vertically and horizontally, with 1.5 m being most common for "drill-and-grout" nails. To enhance satisfactory long-term performance of the reinforced wall, the manual recommends carrying out a ground corrosivity assessment during the site investigation stage and designing appropriate corrosion protection measures according to the assessment results. The eligible corrosion protection measures include grout cover, epoxy coating, or plastic encapsulation. Apparently, use of metallic coating, such as galvanization, is not a common practice in the United States. In comparison with design guidance, advice on construction and maintenance aspects is relatively limited. Guidelines are only given on selected subjects, such as the procurement method, site testing, performance monitoring, and requirements for experienced contractors. The focus of construction guidance is on the application of shotcrete for wall facings.

9.4.2 Geotechnical Engineering Circular No. 7: Soil Nail Walls—Reference Manual

Another well-established guidance document for design and construction of soil nailed walls for highway applications in the United States is Geotechnical Engineering Circular No. 7: Soil Nail Walls—Reference

Manual (Lazarte et al., 2015), which is also published by the Federal Highway Administration of the US Department of Transportation. The guide was first issued in 2003 and was revised and updated subsequently in 2015. It is the latest standard of good practice for application of soil nailing in walls and excavations for roadway projects in the United States. The guide covers permanent structures (50 to 75 years) only, with temporary applications being excluded from the document.

The circular introduces a framework for the design of soil nailed walls taking into account safety factors used in the ASD approach, while integrating the LRDF principles. Similar to the manual by Byrne et al. (1998), the guide recommends the use of limit equilibrium methods in assessing both the potential external and internal stability of a soil nailed wall. For external stability, the circular requires consideration of various failure modes including global failure, sliding, and basal heave. In assessing global stability, the circular suggests use of commercial computer programs to facilitate the analysis in respect of various shapes of potential slip surfaces. For internal stability, the guide depicts a number of potential failure modes that have to be considered, which include pullout failure (bond failure at the interface of soil and grout as well as that of reinforcement and grout), tensile failure of nail reinforcement, wall face bending, and punching shear failure, etc. A set of minimum safety factors for use of the ASD approach, together with load factors and load combinations for the framework of the LRFD approach, are recommended correspondingly by the guide. The results from ASD-based wall stability analysis are utilized in the structural design of soil nails, which is based on LRFD-based load factors. Resistance factors back-calculated from the ASD safety factors for various limit states are also provided in the manual for the LRFD verifications based on the ASD-based wall stability analysis. Pullout resistance is a critical issue in the design. As an initial preliminary estimation, the circular published typical values of bond strength in various soil types using different installation methods. The circular requires validation of the preliminary design value by local experience and site pullout test. Specific guidance on consideration of seismic loads in the design is also provided by the guide. Apart from stability assessment, the guide requires consideration of serviceability, primarily on deformation control. To facilitate deformation analyses, empirical correlations between wall deformation and pertinent parameters, such as nail length, soil type, and wall height, are given by the document.

Regarding the layout and dimensions of soil nails, advice is given on various critical aspects, which include nail distribution pattern, spacing, inclination, and length. Grid pattern (i.e. square or staggered patterns) is suggested to be adopted in conjunction with a spacing of about 1.2 m to 1.8 m, both vertically and horizontally. It is desirable to install nails at about $10°$ to $20°$ (most commonly at $15°$) below horizontal. For nail length, the guide suggests that it should normally fall within 0.5 to 1.2 H (where H is the height of the wall), or otherwise soil conditions have to be re-assessed to check if soil nailing is a suitable engineering scheme.

For durability, in order to select an appropriate level of corrosion protection to the soil nails, the guide recommends carrying out a soil corrosivity assessment to obtain information on the corrosion potential of the ground. Depending on the results of the assessment, various types of corrosion protection measures, including grout cover, sacrificial steel, epoxy coating, galvanizing, and plastic encapsulation, could be selected to suit different levels of ground corrosivity that may be encountered throughout the service life of the structure.

Apart from design aspects, the guide provides comprehensive guidance on various critical construction aspects. These include procurement method, selection of construction materials and methods, qualification requirements for contractors and site supervision personnel, quality control, site testing, and monitoring, etc. Specific guidelines are also given on the contracting approaches and preparation of specifications (method-based and performance-based).

9.5 NORDIC COUNTRIES PRACTICE (DENMARK, FINLAND, ICELAND, NORWAY, AND SWEDEN)

The geotechnical societies in the Nordic countries, including Denmark, Finland, Iceland, Norway, and Sweden, together with the Nordic Industrial Fund, published the first guidance document in 2003, namely Nordic Guidelines for Reinforced Soils and Fills, with the aim to advocate the use of ground reinforcement methods in geotechnical engineering. Although the guidelines are mainly for use of polymeric reinforcement in reinforced fills, the soil nailing technique is also covered by the document. The guide was subsequently revised in 2005 (Rogbeck et al., 2005), which became the standard of good practice in Nordic countries for applying ground reinforcement, including soil nails, for strengthening earth structures. As the development of Eurocodes was still in progress in the early 2000s, the Nordic guidance was primarily formulated on the basis of the European pre-standards (ENV) promulgated at that time (e.g. ENV 1991-1—Basis of Design and Actions of Structures and ENV 1997-1—Geotechnical Design).

The scope of applying the soil nailing technique covered by the guide includes basically excavations and natural slopes, with a service life ranging from less than 6 months (temporary applications) to 100 years (permanent applications). It provides recommendations on the design, construction, method of procurement, supervision, site testing, quality control, and monitoring of soil nailed earth structures. In general, the LSD approach (both ultimate and serviceability limit states) using partial safety factors is adopted in the design, where the partial safety factors are selected corresponding to the geotechnical risk of the structure. The ultimate limit state corresponds to the stability of the soil nailed structure, whereas the serviceability limit state mainly focuses on the deformation of the structure. The guide requires consideration of both external and internal failure modes in assessing the overall stability of a soil nailed structure. The external failure modes include global stability, bearing, sliding, and overturning of the reinforced earth structure. Limit equilibrium methods are recommended for assessing the stability of the nailed structures. Conventional analytical methods, such as the method of slices (e.g. Bishop, Janbu, Morgenstern, & Price methods), are recommended for checking the global stability. For internal stability, the failure modes, including bond failure at the nail-soil interface, nail reinforcement failure due to combined actions of tension, bending and shear, bearing failure in the soil below the nail, and facing failure, should be assessed. To facilitate deformation analysis (i.e. serviceability limit state), the guide provides some empirical correlations between deformation and relevant parameters, such as type of soil, length of soil nails, height of structure, etc. Guidelines on numerical analysis in this aspect are, however, limited. In the estimation of pullout resistance, the guide suggests making reference to the published information from literature (e.g. Clouterre, 1991) as a preliminary design value, followed by verification through field pullout test.

Guidance on the layout and dimensions of soil nails is also given by the document, with reference to the French and UK practice, which are expressed in terms of various ratios between different pertinent parameters, such as length and diameter of soil nails, method of nail installation (i.e. driven or "drill-and-grout"), and height of the earth structure. The guide also suggests installing nails at an angle of 10° to 20° below the horizontal. In fulfilling the durability requirement, the document recommends carrying out a ground corrosivity assessment, so as to classify the corrosive potential of the site. Depending on the results of the assessment and the design life of the soil nailed structure, guidance on selection of appropriate corrosion protection measures to nail reinforcement, including provision of grout cover, sacrificial steel thickness, or plastic encapsulation, is given to suit the ground and environmental conditions.

Apart from design, the guide also provides recommendations and suggestions on various aspects in relation to construction of soil nails. These include method of procurement, selection of construction materials, construction methods, and quality control (i.e. site supervision, testing, and monitoring). Some common problems relating to construction of soil nails and the corresponding preventive and mitigation measures are also provided for reference.

9.6 JAPANESE PRACTICE

The standard of good practice for application of soil nailing currently promulgated in Japan is a guidance document published by the Japan Highway Public Corporation in 1998, namely Design and Construction Guidelines for Soil Nailing (JHPC, 1998). The document is a revised and expanded version of that published in 1987 by the same organization. It provides guidelines on the use of soil nails as temporary (less than 2 years) and permanent engineering measures to reinforce cut slopes and excavations. The scope of the guide includes investigation, design, and construction of soil nailing works. However, the guidelines are mostly applicable for the purpose of serving as a preventive measure against medium-size landslides.

In general, the guide adopts the ASD approach with recommended minimum safety factors against various types of failures. Limit equilibrium methods are used in carrying out stability analyses. In assessing the overall stability of the soil nailed structures, the guide requires consideration of both potential external and internal stability. The external failure modes mainly refer to the global, sliding, bearing, and overturning stability. Method of slices and wedge analysis are suggested for global stability analysis. The internal failure modes correspond to pullout failure (mainly on the bond at the interface between soil and grout as well as between nail reinforcement and grout) and the tensile failure of nail reinforcement. Although the guide also acknowledges a possible need to consider nail reinforcement failure under combined actions of tension, shear, and bending, no guidance is, however, provided on this. Also, little guidance is given by the document on serviceability aspects, such as deformation analysis and requirements. In the estimation of pullout resistance, the guide suggests carrying out the field pullout test in order to determine the design value. Nonetheless, to facilitate preliminary design, notional resistance values related to soil type and SPT N values are given for reference.

Regarding the layout and dimensions of soil nails, the guide recommends that nails should be installed at a spacing of about 1.0 m to 1.5 m, so as to achieve a density of about one nail per 2 m². In order to maximize the mobilization of tensile force, the guide suggests installing nails preferably at an angle between 10° below horizontal and right angle to the slope face. Also, the length of soil nails should normally fall between 2 and 5 m.

For durability, the guide suggests that no corrosion protection to nail reinforcement is necessary if they are for temporary applications. For permanent applications, a ground corrosivity assessment should be carried out to classify the corrosion potential of the site and design the appropriate protection measures. Appropriate corrosion protection measures may include grout cover, sacrificial steel thickness, and galvanization. Alternatively, advanced materials, such as fiber reinforced polymers, could be used as nail reinforcement.

Apart from design aspects, the guide also provides advice on various aspects in relation to construction of soil nails. These include planning and preparation of nail construction, procedures of nail installation, selection of construction plant, site testing, etc. Specific advice is also given on all the critical activities involved in the construction of soil nails. However, little guidance is given on site supervision, including qualification requirements for contractors and site supervision personnel.

9.7 HONG KONG PRACTICE

The soil nailing technique was introduced to Hong Kong in the 1980s. It was first used in Hong Kong as a prescriptive method to provide support to slopes with deeply weathered zones in otherwise sound material. This was followed by a few cases where passive anchors or tie-back systems were used. In the early 1990s, the practices for design of soil nails to improve the stability of existing slopes were summarized by Watkins & Powell (1992), which soon became the norm for soil nail design. With the increasing popularity of the soil nailing technique in the late 1990s, a systematic program of soil nail related studies was undertaken by the Geotechnical Engineering Office (GEO) with the aim to develop a standard of good practice for local practitioners. The studies involved literature review, field load tests, site trials, numerical and physical modeling, and laboratory tests. They have brought about technological advances in respect of design and construction, which resulted in the publication of Geoguide 7—Guide to Soil Nail Design and Construction in 2008 (GEO, 2008). This is the latest guidance document which promulgates state-of-good-practice for the investigation, analysis, design, construction, monitoring, and maintenance of soil nailed structures. The guidance provided by the document is applicable to both permanent (up to 120 years) and temporary (less than 10 years) applications.

Like other international standards and codes, soil nails are designed to improve the stability of earth structures principally through mobilization of tension in the nail reinforcement. The guide adopts primarily the ASD approach for the design of soil nails. Consideration of both the potential external and internal failure modes is required in the course of assessing the overall stability of soil nailed structures. In general, limit equilibrium methods are recommended for the stability assessment. As a minimum, the external failure modes should include global, sliding, and bearing stability. Method of slices is suggested for checking the global stability. Regarding the internal instability, the guide requires examination of various potential failure modes, including pullout failure (bond failure at the interface between ground/grout and reinforcement/grout), tensile failure of nail reinforcement, punching shear failure of nail heads, etc. A set of minimum factors of safety corresponding to the pertinent external and internal failure modes of a soil nailed structure is recommended based on probability of failure, and consequence to life and economic consequence in the event of failure. Apart from stability, the guide requires consideration of serviceability, particularly ground deformation. Deformation analysis may be needed to understand the interaction between soil nails and the ground as well as the effect of deformation on the sensitive facilities in the vicinity of the site. In this regard, the guide provides guidance on the use of numerical modeling for deformation analysis by adopting stress-strain finite element or finite difference methods. In relation to pullout resistance, the guide suggests using the effective stress method to determine the design value and followed by verification using field pullout tests during the construction stage. Alternatively, empirical correlations with soil tests, such as SPT and pressuremeter test, can be adopted to estimate the design value, together with verification by field pullout tests.

Guidelines on the layout and dimensions of soil nails are given by the guide. These include nail distribution pattern, spacing, inclination, and length. Regular square or staggered grid pattern of soil nails with a spacing ranging from 1.5 m to 2 m, both vertically and horizontally, should be adopted. Soil nails should preferably be installed at about 5° to 20° below horizontal. However, if steeply inclined nails are used, the guide also provides a pragmatic approach to assess the reduced effectiveness of the nails. Regarding nail length, the guide suggests avoidance of using long nails (i.e. more than 20 m), or otherwise the potential construction risk associated with this, such as drilling difficulty and excessive grout leakage, should be assessed and mitigation measures provided where necessary. Apart from soil nails, recommendations on the design of nail heads, facing, drainage systems, landscaping, and aesthetics are also provided by the document.

Soil nailed structures should be sufficiently durable and capable of withstanding attack from the existing and envisaged corrosive environment without unduly affecting stability and serviceability. In this regard, the guide recommends carrying out a ground corrosivity assessment and providing appropriate corrosion protection measures. Suitable measures, including grout cover, sacrificial steel thickness, galvanizing or plastic encapsulation, are recommended to suit the different environment and design life of the structure. Alternatively, advanced materials, such as stainless steel and fiber reinforced polymer, can be considered as a replacement for steel reinforcement.

Proper execution and quality control of soil nailing works are essential to the satisfactory completion of construction and long-term performance of soil nailed structures. The guide offers practical advice on various aspects throughout the construction process. These include pre-construction review, setting out, drilling, nail installation, excavation, site supervision, etc. Specific guidance is also given on quality control and site testing, such as destructive and non-destructive test, and field pullout test. In addition, the guide highlights some common construction problems that may be encountered and the corresponding preventive or mitigation measures. The guidelines also cover monitoring and maintenance issues. In conjunction with the guide, a set of specifications and standard construction drawings on soil nailing works is also available for use by the geotechnical practitioners (CEDD, 2006).

Table 9.2 Key characteristics of standards of good practice of soil nailing in different places

	France	UK	United States	Nordic Countries	Japan	Hong Kong
Standards/Codes of Practice	Clouterre (1991)	DOT (1994); Phear et al. (2005); BSI (2010, 2013)v; (O'Donovan et al., 2020)	Byrne et al. (1998); Lazarte et al. (2015)	Rogbeck et al. (2005)	JHPC (1998)	GEO (2008)
Scope of Application	Retaining walls and excavations	Slopes, retaining walls, embankments, and excavations (guidance basically applicable to "drill-and-grout" soil nails)	Retaining walls and excavations (guidance basically applicable to "drill-and-grout" soil nails)	Slopes and excavations	Cut slopes and excavations (only address medium size failure)	Slopes, retaining walls, and excavations (guidance basically applicable to "drill-and-grout" soil nails)
Design Life	Up to 100 years	Up to 120 years (up to 60 years in DOT, 1994)	Up to 100 years (up to 75 years and excludes temporary applications in Lazarte et al., 2015)	Up to 100 years	Less than 2 years to permanent (period not specified)	Up to 120 years
Design Approach	LSD with partial safety factors	LSD with partial factors of safety	ASD with integration of LSD principles (ASD is adopted in stability analysis while LSD is used in soil nail design)	LSD with partial factors of safety	ASD	ASD

(Continued)

Table 9.2 (Continued) Key characteristics of standards of good practice of soil nailing in different places

	France	UK	United States	Nordic Countries	Japan	Hong Kong
Failure Modes	External and internal (including combined actions of tension, bending, and shear in soil nails)	External and internal	External and internal	External and internal (including combined actions of tension, bending, and shear in soil nails)	External and internal	External and internal
Method of Analysis	Limit equilibrium (method of slices)	Limit equilibrium (method of slices and two-part wedge method)	Limit equilibrium (method of slices and two-part wedge method)	Limit equilibrium (method of slices)	Limit equilibrium (method of slices)	Limit equilibrium (method of slices)
Deformation Analysis	Empirical correlation with soil type, nail and wall geometry	Limited guidance	Empirical correlation with soil type, nail and wall geometry	Empirical correlation with soil type, nail and wall geometry	Limited guidance	Guidance on numerical analysis

(Continued)

Table 9.2 (Continued) Key characteristics of standards of good practice of soil nailing in different places

	France	UK	United States	Nordic Countries	Japan	Hong Kong
Estimation of Pullout Resistance	Empirical correlation with soil type and soil tests, such as pressuremeter test; field pullout test	Empirical correlation with soil tests, such as SPT and pressuremeter test; published data; analytical methods, such as effective stress method and total stress method; field pullout test	Empirical correlation with soil tests, such as SPT; published data; local experience; field pullout test	Published data and field pullout test	Published data and field pullout test	Effective stress method; empirical correlation with soil tests, such as SPT and pressuremeter test; field pullout test
Durability	Site aggressivity assessment and appropriate corrosion protection measures to reinforcement (sacrificial steel thickness, metallic and non-metallic coating, and plastic encapsulation)	Site aggressivity assessment and appropriate corrosion protection measures to reinforcement (grout cover, sacrificial steel thickness, metallic and non-metallic coating, and plastic encapsulation) or other corrosion-resistant materials (stainless steel and FRP)	Site aggressivity assessment and appropriate corrosion protection measures to reinforcement (grout cover, sacrificial steel thickness, metallic and non-metallic coating, and plastic encapsulation)	Site aggressivity assessment and appropriate corrosion protection measures to reinforcement (grout cover, sacrificial steel thickness, metallic and non-metallic coating, and plastic encapsulation)	Site aggressivity assessment and appropriate corrosion protection measures to reinforcement (sacrificial steel thickness, metallic and non-metallic coating) or FRP reinforcement	Site aggressivity assessment and appropriate corrosion protection measures to reinforcement (grout cover, sacrificial steel thickness, metallic and non-metallic coating, and plastic encapsulation) or other corrosion-resistant materials (stainless steel and FRP)

(Continued)

Table 9.2 (Continued) Key characteristics of standards of good practice of soil nailing in different places

		France	UK	United States	Nordic Countries	Japan	Hong Kong
Layout and Dimensions	Pattern	Regular	Regular square or staggered grid	Regular square or staggered grid	No guidance	No guidance	Regular square or staggered grid
	Inclination	5° to 15° below horizontal	5° to 10° below horizontal	5° to 20° (most commonly at 15°) below horizontal	10° to 20° below horizontal	10° below horizontal to right angle to slope face	5° to 20° below horizontal
	Spacing/Density	About 0.7 m to 1 m (1 to 2 nails/m²) for driven nails About 1.5 m to 2.5 m (0.15 to 0.4 nails/m²) for "drill-and-grout" nails	1.5 m to 3 m (vertical and horizontal for gentle slopes) 1 m to 2 m (vertical and horizontal for intermediate slopes) 0.75 m to 1.5 m (vertical for steep slopes) 0.5 m to 2 m (horizontal for steep slopes)	1 m to 2 m (vertical and horizontal) with 1.5 m being most common	Various ratios between geometry of wall, nail installation method, soil type, and geometry of nail (hole diameter, nail spacing and length, etc.), based on practice in France and the UK, are given	1 m to 1.5 m (1 nail/2 m²)	1.5 m to 2 m
	Length	0.5 to 0.7 of slope/wall height (H) for driven nails 0.8 H to 1.2 H for "drill and grout" nails	0.5 H to 2 H for gentle slopes 0.5 H to 1.5 H for intermediate slopes 0.5 H to 1.2 H for steep slopes	0.5 H to 1.2 H		Normally 2 m to 5 m	Preferably less than 20 m

(Continued)

Table 9.2 (Continued) Key characteristics of standards of good practice of soil nailing in different places

	France	UK	United States	Nordic Countries	Japan	Hong Kong
Facing	Shotcrete and reinforced concrete	Non-structural soft for gentle slopes Structural flexible for intermediate slopes Structural hard for steep slopes	Shotcrete and reinforced concrete	No facing may be necessary for slope angle less than 30° Soft or flexible may be used with consideration of ground deformation Hard facing for steep slopes	Use of soft, flexible, and hard facing for different slope setting	Use of soft, flexible, and hard facing for different range of slope angle. Guidance on suitable local planting species is also given
Construction	Installation method, selection of equipment, specification, and quality control	Percussive, vibratory, jacked, fired and drill-and-grout methods	Guidance mainly for shotcrete wall facing	Guidance is given on various construction aspects including procurement method, selection of materials, construction process, site supervision, quality control, testing, and monitoring	Guidance is given for planning and preparation of construction, procedures of nail installation, selection of construction plant, site testing, etc.	Guidance is given on various construction aspects, such as pre-construction review, setting out, selection of materials, drilling, nail installation, site supervision, quality control, and testing

(Continued)

Table 9.2 (Continued) Key characteristics of standards of good practice of soil nailing in different places

	France	UK	United States	Nordic Countries	Japan	Hong Kong
Remarks			Guidelines on frost protection, handling of seismic loads, and expansive soils are provided			

Geoguide 7 provides comprehensive guidance on the design and construction of soil nailed structures, where analytical approach has been adopted in the design process. As an alternative to the analytical approach, the GEO published another guidance document, namely GEO Publication 1/2009: Prescriptive Measures for Man-made Slopes and Retaining Walls (GEO, 2009), which promulgates an experience-based method for the design of slopes and retaining walls using soil nails. Standard layout and dimensions of soil nails can be selected from the publication to suit the environment and geometry of the earth structures as well as the level of safety improvement to be designed. This prescriptive approach allows the design of soil nails in situations where conventional calculation methods are not necessary or not available.

9.8 COMPARISON OF INTERNATIONAL STANDARDS

The principle of the soil nailing technique is to reinforce earth structures by mobilizing tensile forces in soil nails effectively. Since the 1970s, different parts of the world have carried out technical development work on the technique and developed their own standards and codes of practice for temporary and permanent applications. Sections 9.2 to 9.7 above briefly illustrate the key characteristics of state-of-good-practice of soil nailing works being promulgated in North America, Europe, and Asia. One may recognize that while these standards and codes share many similarities, some differences are also noted, such as the scope of application, design approach, and the methodology for estimating pullout resistance. Table 9.2 summarizes the key characteristics of the pertinent standards and codes of practice.

9.9 CONCLUDING REMARKS

Since the surge in popularity of using the soil nailing technique in the 1970s, different places have carried out soil nail related research and development work and published their own local standards to promulgate the good practice for reinforcing earth structures. These standards have emerged primarily from the specific needs of infrastructure development, and the enhanced understanding and extensive experience of soil nailing. In this chapter, state-of-good-practice for applying the soil nailing technique in France, the United States, the UK, the Nordic Countries, Japan, and Hong Kong has been examined and compared. Given its good track record in performance in various engineering applications, soil nailing has been proven as a cost-effective engineering solution to stabilize slopes, retaining walls, embankments, and excavations. Although there are some subtle variations amongst the practice being promulgated in different places, the key principles of the soil nailing technique are universally applicable.

REFERENCES

BSI (British Standards Institution) (2010). Execution of Special Geotechnical Works: Soil Nailing (BS EN 14490: 2010). British Standards Institution, London, UK, 68 p.

BSI (2013). Code of Practice for Strengthened/Reinforced Soils—Part 2: Soil Nail Design (BS 8006-2: 2011). British Standards Institution, London, UK, 104 p.

Byrne, R.J., Cotton, D., Porterfield, J., Wolschlag, C. & Ueblacker, G. (1998). Manual for Design and Construction Monitoring of Soil Nail Walls. Federal Highway Administration, US Department of Transportation, Washington, DC, USA, Report No. FHWA-SA-96-069R, 530 p.

CEDD (Civil Engineering and Development Department) (2006). *General Specification for Civil Engineering Works*. Civil Engineering and Development Department, HKSAR Government, Hong Kong.

Clouterre (1991). French National Research Project Clouterre: Recommendations Clouterre (English Translation 1993). Federal Highway Administration, US Department of Transportation, Washington, DC, USA, Report No. FHWA-SA-93-026, 321 p.

DOT (Department of Transport) (1994). Design Methods for the Reinforcement of Highway Slopes by Reinforced Soil and Soil Nailing Techniques. In *Design Manual for Roads and Bridges, Part 4 (HA 68/94)*. The Highways Agency, The Department of Transport, Washington, DC, UK, 112 p.

GEO (Geotechnical Engineering Office) (2008). *Guide to Soil Nail Design and Construction (Geoguide 7)*. Geotechnical Engineering Office, Civil Engineering and Development Department, HKSAR Government, Hong Kong, 97 p.

GEO (2009). *Prescriptive Measures for Man-made Slopes and Retaining Walls (GEO Publication No. 1/2009)*. Geotechnical Engineering Office, Civil Engineering and Development Department, HKSAR Government, Hong Kong, 74 p.

JHPC (Japan Highway Public Corporation) (1998). *Design and Construction Guidelines for Soil Nailing*. Japan Highway Public Corporation, Japan, 111 p (in Japanese).

Lazarte, C.A., Robinson, H., Gomez, J.E., Baxter, A., Cadden, A. & Berg, R. (2015) Geotechnical Engineering Circular No. 7: Soil Nail Walls—Reference Manual. Federal Highway Administration, US Department of Transportation, Washington, DC, USA, Report No. FHWA-NHI-14-007, 385 p.

O'Donovan, J., Feudale Foti, E., Turner, M., Irvin, C. & Mothersille, D. (2020). *Grouted Anchors and Soil Nails - Inspection, Condition Assessment and Remediation*. Construction Industry Research & Information Association, London, UK, CIRIA Report No. C794, 193 p.

Phear, A., Dew, C., Ozsoy, B., Wharmby, N.J., Judge, J. & Barley, A.D. (2005). Soil Nailing: Best Practice Guidance. Construction Industry Research & Information Association, London, UK, CIRIA Report No. C637, 286 p.

Rogbeck, Y., Alen, C., Franzen, G., Kjeld, A., Oden, K., Rathmayer, H., Watn, A. & Oiseth, E. (2005). *Nordic Guidelines for Reinforced Soils and Fills*. Geotechnical Societies of Denmark, Finland, Norway and Sweden, Nordic Geosynthetic Group, Rev. B.

Watkins, A.T. & Powell, G.E. (1992). Soil nailing to existing slopes as landslip preventive works. *Hong Kong Engineer*, vol. 20, pp. 20–27.

Appendix A
Worked example (Allowable Stress Design)

STEP 1—ESTABLISH DESIGN REQUIREMENTS

- Soil nails are proposed to stabilize an existing soil cut slope at 60° to the horizontal with a maximum slope height of up to 9 m. The slope crest is a footpath and a natural slope and the slope toe is a two-lane-two-way carriageway with medium dense traffic flow. The geological material comprises decomposed volcanic rock, i.e. completely to highly decomposed tuff (Figure A.1).
- The design life of the slope is 120 years.
- Adequate safety margin should be provided to the soil nailed slope against all the perceived potential modes of failure throughout its service life.

STEP 2—IDENTIFY MODES OF FAILURE

- Both potential external and internal modes of failure are identified. The potential external failure mode includes overall failure, and the internal failure modes include (i) tensile failure of soil nail reinforcement, (ii) pullout failure at soil-grout interface, (iii) pullout failure at grout-reinforcement interface, and (iv) bearing failure underneath soil nail heads.

STEP 3—IDENTIFY LOADING CONDITIONS

- A groundwater table in response to a 10-year return period rainfall event is considered.
- A footpath is located along the slope crest. A 5 kPa uniformly distributed surcharge is applied at the crest of the slope to cater for the loading on the footpath.

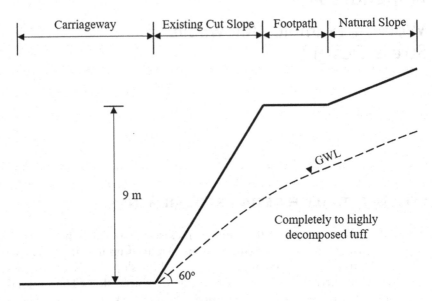

Figure A.1 Cross-section of the cut slope.

STEP 4—IDENTIFY SITE CONSTRAINTS

- It is inevitable that construction works will be carried out in close proximity to the carriageway in front of the slope. A temporary traffic arrangement for lane closure is required to erect working platforms for soil nailing works.

STEP 5—DESIGN PRELIMINARY LAYOUT AND DIMENSIONS OF SOIL NAILS

- It is proposed that 10 m long soil nails comprising 25 mm diameter steel bars be adopted. "Drill-and-grout" method is used to construct 150 mm diameter soil nails.
- Soil nails installed at vertical and horizontal spacing of 1.5 m staggered in a triangular pattern are proposed, i.e. six rows of soil nails as shown in Figure A.2.

STEP 6—DESIGN CODE AND DESIGN METHOD

- The Allowable Stress Design (ASD) approach using Geoguide 7 (GEO, 2008) is adopted.

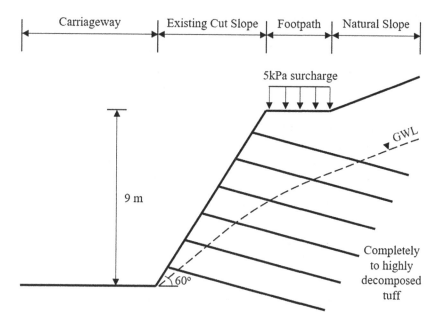

Figure A.2 Preliminary layout of soil nails.

Table A.1 Required minimum factor of safety against external and internal failure modes

Mode of Failure	Minimum Factor of Safety
Overall	$F_G = 1.4$
Tensile failure of soil nail reinforcement	$F_T = 1.5$
Pullout failure at soil-grout interface	$F_P = 1.5$
Pullout failure at grout-reinforcement interface	$F_R = 2.0$
Bearing failure of nail head	$F_B = 3.0$

- Based on the consequence-to-life and economic consequence of the slope, the required minimum factors of safety against potential external and internal modes of failure are given in Table A.1.

STEP 7—ESTABLISH GEOMETRY AND DESIGN CROSS-SECTION

- As the slope geometry varies significantly along the slope, several cross-sections should be considered in the analysis. This example considers the cross-section with the maximum height as the critical section for the soil nail design (Figure A.1).

Table A.2 Soil parameters adopted in the design

Soil Stratum	Unit Weight (kN/m³)	c' (kPa)	φ' (Degree)	Soil Aggressivity
Completely to highly decomposed tuff	19	2	35	Aggressive

STEP 8—ESTABLISH GROUND MODEL AND SOIL PARAMETERS

- Based on the results of site-specific ground investigation, a ground model is established, which comprises a single stratum of completely to highly decomposed tuff with no adverse geological features. Laboratory tests including soil classification test, soil aggressivity test, and consolidated undrained triaxial test are carried out to obtain the soil parameters for the design (Table A.2).

STEP 9—ESTABLISH GROUNDWATER MODEL

- Reference has been made to local practice and the groundwater table is assumed to be about two-thirds of the slope height based on the groundwater monitoring record to account for a 10-year return period rainfall event (Figure A.1).

STEP 10—METHOD OF ANALYSIS

- Equilibrium analysis coupled with method of slices that satisfies both force and moment equilibrium (Morgenstern-Price method) is adopted.

STEP 11—EXTERNAL STABILITY CHECK

- The slope stability computer program using the Morgenstern-Price method is adopted to determine the minimum factor of safety against overall stability of the existing slope and the proposed nailed slope. The factor of safety of the existing slope is found to be 1.03 (Figure A.3).
- It is found that the required minimum safety factor against overall failure, F_G, of 1.2 can be achieved based on the preliminary design results, i.e. six rows of soil nails at vertical and horizontal spacing of 1.5 m in a staggered pattern (Figure A.4). The corresponding required nail forces are shown in Table A.3.

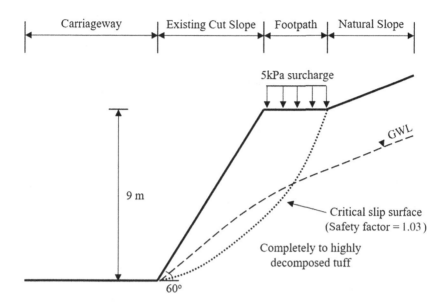

Figure A.3 Stability analysis of the existing cut slope.

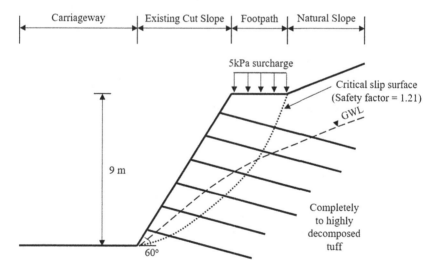

Figure A.4 Stability analysis of the proposed nailed slope.

Table A.3 Required nail forces to achieve the minimum factor of safety

Row No.	Required Soil Nail Force (kN/m)	Required Soil Nail Force (kN Per Nail)
I (Top)	40.0	60.0
2	40.0	60.0
3	40.0	60.0
4	45.0	67.5
5	45.0	67.5
6 (Bottom)	45.0	67.5

STEP 12—INTERNAL STABILITY CHECK

- The internal stability checks of a soil nail include (i) tensile failure of soil nail reinforcement, (ii) pullout failure at soil-grout interface, (iii) pullout failure at grout-reinforcement interface, and (iv) bearing failure underneath soil nail heads.
- The pullout resistance of a soil nail is the smallest of the allowable tensile force of soil nail reinforcement, allowable pullout resistance at soil-grout interface, and allowable pullout resistance at grout-reinforcement interface.

(a) Allowable tensile force of soil nail reinforcement is determined by:

$$T_T = (D/2)2\pi f_y / F_T$$

where f_y is the yield strength of soil nail reinforcement

D is the diameter of soil nail

F_T is the factor of safety against tensile failure of soil nail reinforcement (= 1.5)

(b) Allowable pullout resistance at soil-grout interface is determined by:

$$T_P = [c'P_c + 2D\sigma'_v\mu^*]L / F_P$$

where c' is the effective cohesion of the soil

ϕ' is the angle of shearing resistance of the soil

μ^* is the apparent coefficient of friction between the soil nail surface and the soil (= $\tan\phi'$)

σ'_v is the effective vertical stress acting on the soil nail

L is the bond length of the soil nail in the passive zone

P_c is the perimeter of the soil nail

D is the diameter of the soil nail

F_p is the factor of safety against pullout failure at soil-grout interface (= 1.5)

(c)   Allowable pullout resistance at grout-reinforcement interface is determined by:

$$T_R = \frac{\beta\sqrt{f_{cu}P_rL}}{F_R}$$

where β is the coefficient of friction at the grout-reinforcement interface, which depends on the bar type characteristic

f_{cu} is the characteristic strength of cement grout

P_r is the effective perimeter of the soil-nail reinforcement

F_R is the factor of safety against pullout failure at grout-reinforcement interface (= 2.0)

- To achieve the required safety factors and provide the required pullout resistance of the soil nails, the design length and diameter of soil nails are summarized in Table A.4.
- For checking the bearing failure of soil underneath nail head, reference has been made to Table 4.17. Given that the slope angle, effective cohesion, and angle of shearing resistance of the soil are 60°, 4 kPa, and 36°, respectively, square soil nail head of size 600 mm × 600 mm × 250 mm (thick) is proposed.

STEP 13—SERVICEABILITY

- As the slope is an existing cut slope and there is no sensitive receiver at the slope crest and toe, excessive deformation of the soil nailed slope is not a concern. Therefore, a detailed deformation analysis is considered not necessary.

Table A.4 Required nail lengths to achieve the minimum factor of safety

| Row No. | Soil Nail Length (m) | Diameter (mm) | | Pullout Resistance (kN Per Nail) | Required Soil Nail Force (kN Per Nail) |
		Steel Bar	Drillhole		
I (Top)	10	25	150	60.2	60.0
2	10	25	150	66.8	60.0
3	9	25	150	69.9	60.0
4	9	25	150	87.4	67.5
5	8	25	150	103.9	67.5
6 (Bottom)	8	25	150	111.9	67.5

STEP 14—DURABILITY AND CORROSION PROTECTION MEASURES

- Soil aggressivity assessment is carried out based on soil composition and chemical test results, and the soil aggressivity is classified as "Aggressive" (Table A.2). Class 1 corrosion protection measures (i.e. zinc galvanization of a minimum of 610 g/m^2 and corrugated plastic sheathing) for the soil nails are adopted (reference to Tables 5.14 and 5.15).

STEP 15—SURFACE DRAINAGE SYSTEM

- An existing surface drainage system was observed during site inspections. As its current condition is good and its capacity is sufficient, it is proposed to retain the existing drainage system.

STEP 16—AESTHETICS AND LANDSCAPING

- Given that the slope angle is 60°, 75 mm thick sprayed concrete with planter holes for climbers is recommended as the surface protection to prevent from surface erosion and enhance aesthetics of the slope (reference to Figure 4.16).

REFERENCE

GEO (Geotechnical Engineering Office) (2008). *Guide to Soil Nail Design and Construction (Geoguide 7)*. Geotechnical Engineering Office, Civil Engineering and Development Department, HKSAR Government, Hong Kong, 97 p.

Appendix B
Worked example (Limit State Design)

STEP I—ESTABLISH DESIGN REQUIREMENTS

- Soil nails are proposed to stabilize an existing soil cut slope at 60° to the horizontal with a maximum slope height of up to 9 m. The slope crest is a footpath and a natural slope, and the slope toe is a two-lane-two-way carriageway with medium dense traffic flow. The geological material comprises decomposed volcanic rock, i.e. completely to highly decomposed tuff (Figure A.5).
- The design life of the slope is 120 years.
- Adequate safety margin should be provided to the soil nailed slope against all the perceived potential modes of failure throughout its service life.

STEP 2—IDENTIFY MODES OF FAILURE

- Both potential external and internal modes of failure are identified. The potential external failure mode includes overall failure, and the internal failure modes include (i) tensile failure of soil nail reinforcement, (ii) pullout failure at soil-grout interface, and (iii) bearing failure underneath soil nail heads.

STEP 3—IDENTIFY LOADING CONDITIONS

- A worst credible groundwater table perceived over the design life of the slope under normal operations is considered.
- A footpath is located along the slope crest. A 5 kPa uniformly distributed surcharge is applied at the crest of the slope to cater for the loading on the footpath.

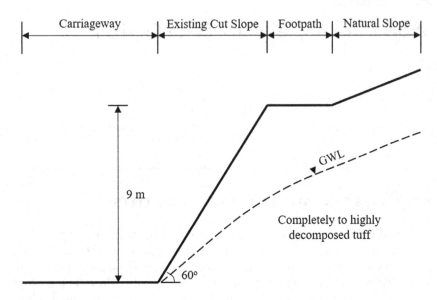

Figure A.5 Cross-section of the cut slope.

STEP 4—IDENTIFY SITE CONSTRAINTS

- It is inevitable that construction works will be carried out in close proximity to the carriageway in front of the slope. A temporary traffic arrangement for lane closure is required to erect working platforms for soil nailing works.

STEP 5—DESIGN PRELIMINARY LAYOUT AND DIMENSIONS OF SOIL NAILS

- It is proposed that 11 m long soil nails (i.e. 1.2 times of slope height) comprising 25 mm diameter steel bars be adopted. "Drill-and-grout" method is used to construct 150 diameter soil nails.
- Soil nails installed at vertical and horizontal spacing of 1.5 m staggered in a triangular pattern are proposed, i.e. six rows of soil nails as shown in Figure A.6.

STEP 6—DESIGN CODE AND DESIGN METHOD

- The Limit State Design (LSD) approach using BS 8006-2:2011 (BSI, 2013) is adopted.
- Two sets of partial factors are applied to load and resistance for checking of ultimate limit state, as summarized in Table A.5.

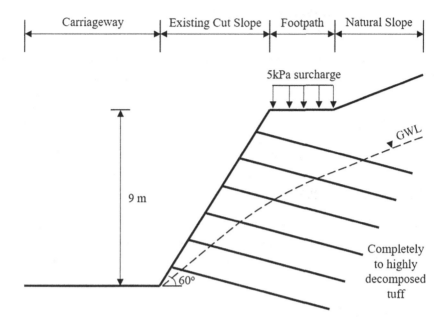

Figure A.6 Preliminary layout of soil nails.

STEP 7—ESTABLISH GEOMETRY AND DESIGN CROSS-SECTION

- As the slope geometry varies significantly along the slope, several cross-sections should be considered in the analysis. This example considers the cross-section with the maximum height as the critical section for the soil nail design (Figure A.5).

Table A.5 Partial factors to load and resistance

Description of Load/Resistance		Partial Factors	
		Set 1	Set 2
Action	Self-weight of soil, W	$\gamma_g = 1.35$	$\gamma_g = 1.0$
	Variable surcharge, q_v	$\gamma_{qv} = 1.5$	$\gamma_{qv} = 1.3$
	Groundwater pressure, u	$\gamma_u = 1.0$	$\gamma_u = 1.0$
Material Properties	Angle of shearing resistance, tan ϕ'	$\gamma_{\tan \phi'} = 1.0$	$\gamma_{\tan \phi'} = 1.3$
	Effective cohesion, c'	$\gamma_{c'} = 1.0$	$\gamma_{c'} = 1.3$
Soil Nail Resistances	Bond stress, τ_b (*Effective Stress*)	$\gamma_{\tau b} = 1.1$	$\gamma_{\tau b} = 1.5$
	Tendon strength, T	$\gamma_s = 1.0$	$\gamma_s = 1.15$ for steel

STEP 8—ESTABLISH GROUND MODEL AND SOIL PARAMETERS

- Based on the results of site-specific ground investigation, a ground model is established, which comprises a single stratum of completely to highly decomposed tuff with no adverse geological features. Laboratory tests including soil classification test, soil aggressivity test, and consolidated undrained triaxial test are carried out to obtain the soil parameters for the design. The soil parameters with applied partial factors (shown in bracket) for the design are shown in Table A.6.

STEP 9—ESTABLISH GROUNDWATER MODEL

- The groundwater table is assumed to be about two-thirds of the slope height based on the groundwater monitoring record to account for the worst credible situation under normal operations over the service life (Figure A.5).

STEP 10—METHOD OF ANALYSIS

- Equilibrium analysis coupled with method of slices (Bishop's simplified method) is adopted.

STEP 11—EXTERNAL STABILITY CHECK

- The slope stability computer program using Bishop's simplified method is adopted to check against overall failure of the existing slope and the proposed nailed slope (Figure A.7).
- Driving moment, $M_{driving}$, and resisting moment, $M_{resisting}$, at the center of rotation using factored values for each parameter are calculated. The value of the model factor, γ_{Sd}, is taken to be unity. The soil nail design is carried out through an iterative process of checking and refinement until for all possible slips using the two sets of partial factors that:

$$\gamma_{Sd} M_{driving} \leq M_{resisting}$$

Table A.6 Soil parameters adopted in the design

Set of Partial Factors	Unit Weight (kN/m³)	c_k' (kPa)	ϕ_k' (Degree)
Set 1	25.7 (1.35)	2.0 (1.0)	35.0 (1.0)
Set 2	19.0 (1.0)	1.5 (1.3)	28.3 (1.3)

Note: Partial factors shown in brackets.

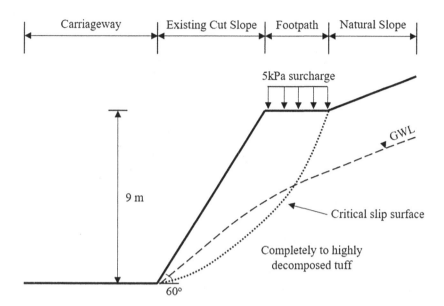

Figure A.7 Stability analysis of the existing cut slope.

- It is found that the checking against ultimate limit state can be fulfilled based on the preliminary design results (Figure A.8). The corresponding required nail forces are shown in Table A.7.

STEP 12—INTERNAL STABILITY CHECK

- The internal stability checks of a soil nail include (i) tensile failure of soil nail reinforcement, (ii) pullout failure at soil-grout interface, and (iii) bearing failure underneath soil nail heads.
- The pullout resistance of a soil nail is the smallest of the design tensile force of soil nail reinforcement and design pullout resistance at soil-grout interface.
- (a) Design tensile force of soil nail reinforcement is determined by:

$$R_{td} = A_{s,nom} f_{yk} / \gamma_s$$

where $A_{s,nom}$ is the nominal cross-sectional area of the soil nail reinforcement

f_{yk} is the characteristic yield strength of soil nail reinforcement

(b) Design pullout resistance at soil-grout interface is determined by:

$$T_d = \lambda_f k_r \sigma_v ' \tan \phi_k ' P_c L / \gamma_{\tau b}$$

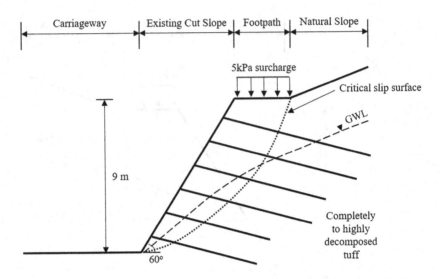

Figure A.8 Stability analysis of the proposed nailed slope.

Table A.7 Required nail forces to fulfill checking of ultimate limit state

Row No.	Required Soil Nail Force (kN/m)	Required Soil Nail Force (kN Per Nail)
I (Top)	50.0	75.0
2	50.0	75.0
3	50.0	75.0
4	55.0	82.5
5	55.0	82.5
6 (Bottom)	55.0	82.5

where λ_f is the interface factor, which is dependent upon the nail installation method

k_r is the factor relating the average radial effective stress around
the nail to the vertical

σ_v' is the effective vertical stress acting on the soil nail

ϕ_k' is the factored angle of shearing resistance of the soil

P_c is the perimeter of the soil nail

L is the bond length of the soil nail in the passive zone

- To provide the required pullout resistance of the soil nails for the ultimate limit state, the design length and diameter of soil nails are summarized in Table A.8.
- The required size of soil nail head is determined by (reference to
Figure 4.14):

Table A.8 Required nail lengths to fulfill checking of ultimate limit state

| Row No. | Soil Nail Length (m) | Diameter (mm) | | Pullout Resistance (kN Per Nail) | Required Soil Nail Force (kN Per Nail) |
		Steel Bar	Drillhole		
I (Top)	14	25	150	78.9	75.0
2	14	25	150	83.9	75.0
3	13	25	150	77.8	75.0
4	13	25	150	91.4	82.5
5	11	25	150	99.5	82.5
6 (Bottom)	11	25	150	141.7	82.5

$$w = \sqrt[3]{\frac{T}{\eta}}$$

$$\eta = \frac{\gamma(1 - r_u)\tan\beta_s e^{3\left(\frac{\pi}{4} - \frac{\phi'}{2} + \alpha_s\right)\tan\phi'}}{2\cos\left(\frac{\pi}{4} + \frac{\phi'}{2}\right)(1 - \sin\phi')}$$

where T is the design load of soil nail
γ is the unit weight of soil
r_u is the pore pressure parameter
β_s is the slope angle
ϕ' is the angle of shear resistance of soil
α_s is the inclination of soil nail

- Square soil nail head of size 600 mm × 600 mm × 250 mm (thick) is proposed.

STEP 13—SERVICEABILITY

- As the slope is an existing cut slope and there is no sensitive receiver at the slope crest and toe, excessive deformation of the soil nailed slope is not a concern. Therefore, a detailed deformation analysis is considered not necessary.

STEP 14—DURABILITY AND CORROSION PROTECTION MEASURES

- An assessment of degradation risk for soil nails is carried out using the method presented in BS EN 14490:2010 (BSI, 2010). The assessment suggests that the soil is "Corrosive." As the risk category is medium, coated steel surrounded by grouted impermeable ducting for the soil nails is recommended (Table 5.10).

STEP 15—SURFACE DRAINAGE SYSTEM

- An existing surface drainage system was observed during site inspections. As its current condition is good and its capacity is sufficient, it is proposed to retain the existing drainage system.

STEP 16—AESTHETICS AND LANDSCAPING

- Given that the slope angle is 60°, shotcrete facing is recommended as the surface protection to prevent from surface erosion.

REFERENCES

BSI (British Standards Institution) (2010). Execution of Special Geotechnical Works—Soil Nailing (BS EN 14490: 2010). British Standards Institution, London, UK, 68 p.

BSI (British Standards Institution) (2013). Code of Practice for Strengthened/Reinforced Soils. Part 2: Soil Nail Design (BS 8006: Part 2: 2011). British Standards Institution, London, UK, 104 p.

Index

Printed in the United States
By Bookmasters